KU-707-183

University of Liverpool

Withdrawn from stock

The Rights of Patients

AN AMERICAN CIVIL LIBERTIES UNION HANDBOOK

THE RIGHTS
OF PATIENTS

THE BASIC ACLU GUIDE TO PATIENT RIGHTS

SECOND EDITION
Completely Revised and Up-to-Date

by

George J. Annas

Springer Science+Business Media, LLC

Copyright © 1989, 1992 by George J. Annas and the American Civil Liberties Union.

Originally published by Humana Press in 1989,1992
Softcover reprint of the hardcover 2nd edition 1992

All rights reserved.

Library of Congress Cataloging-in-Publication Data

Annas, George J.
    The rights of patients: the basic ACLU guide to patient rights

George J. Annas.—2d ed., completely rev. and up-to-date.
        p. cm.—(An American Civil Liberties Union handbook)
        Rev. ed. of: The rights of hospital patients, © 1975.

        Includes bibliographical references and index.
        1. Hospital patients—Legal status, laws, etc.—United States.
        2. Hospitals—Law and legislation—United States. I. Annas, George.
            J. Rights of hospital patients. II. Title. III. Series.

KF3823.A96 1989
344. 73'03211—dc19 [347.3043211]   88-29893
ISBN 978-1-4612-6743-0   ISBN 978-1-4612-0397-1 (eBook)   CIP
DOI 10.1007/978-1-4612-0397-1
This publication is printed on acid-free paper. ∞

*To the sick, and the healers
who treat them as persons*

Contents

Appendixes

Preface

This guide sets forth your rights under the present law and offers suggestions on how they can be protected. It is one of a continuing series of handbooks published in cooperation with the American Civil Liberties Union (ACLU).

Surrounding these publications is the hope that Americans, informed of their rights, will be encouraged to exercise them. Through their exercise, rights are given life. If they are rarely used, they may be forgotten, and violations may become routine.

This guide offers no assurances that your rights will be respected. The laws may change, and in some of the subjects covered in these pages, they change quite rapidly. An effort has been made to note those parts of the law where movement is taking place, but it is not always possible to predict accurately when the law *will* change.

Even if the laws remain the same, their interpretations by courts and administrative officials often vary. In a federal system such as ours, there is a built-in problem since state and federal law differ, not to speak of the confusion between states. In addition, there are wide variations in the ways in which particular courts and administrative officials will interpret the same law at any given moment.

If you encounter what you consider to be a specific abuse of your rights, you should seek legal assistance. There are a number of agencies that may help you, among them ACLU affiliate offices, but bear in mind that the ACLU is a limited-purpose organization. In many communities, there are federally funded legal service offices that provide assistance to persons who cannot afford the costs of legal representation. In general, the rights that the ACLU defends are freedom of inquiry and expression; due process of law; equal protection of the laws; and privacy. The authors in this series have discussed other rights (even though they sometimes fall outside the ACLU's usual concern) in order to provide as much guidance as possible.

These books have been planned as guides for the people directly affected: thus the question and answer format. (In some areas, there are more detailed works available for experts.) These guides seek to raise the major issues and inform the nonspecialist of the basic law on the subject. The authors of these books are themselves specialists who understand the need for information at "street level."

If you encounter a specific legal problem in an area discussed in one of these handbooks, show the book to your attorney. Of course, he or she will not be able to rely exclusively on the handbook to provide you with adequate representation. But if your attorney hasn't had a great deal of experience in the specific area, the handbook can provide helpful suggestions on how to proceed.

Norman Dorsen, Past President
American Civil Liberties Union

The principal purpose of this handbook, as well as others in this series, is to inform individuals of their legal rights. The authors from time to time suggest what the law should be, but their personal views are not necessarily those of the ACLU. For the ACLU's position on the issues discussed in this handbook the reader should write to Librarian, ACLU, 132 West 43d Street, New York, NY 10036.

Acknowledgments

I am pleased to acknowledge the insights of my law, medicine, and public health students over the past decade and a half; many of them will find their questions reflected in this book. My health law colleagues, including Leonard Glantz, Barbara Katz, Wendy Mariner, Frances Miller, John Robertson, and Ken Wing, have consistently been generous with their comments and critiques over the years. Special thanks are due to Joseph M. Healey, who was kind enough to again review the entire manuscript and who is responsible for many of the ideas in this book. Michael Grodin and John Tobias Nagurney also made constructive suggestions. I was helped along the way by dozens of research assistants, all of whom I am indebted to. Those who had the most influence over the final text include Scott Berman, Joan Densberger, Sherry Liebowitz, Martha Neary, and Elizabeth Truesdell. Mary Lou Hannigan, assisted by Trudy David, deserves more of an acknowledgment than these words convey for putting the final manuscript in useful form. All of the physicians, lawyers, and judges I have had the privilege to teach the issues addressed in this book have all contributed to my thinking on patient rights through their questions and comments. The consistent support of my work on patient rights over the past decade by the Boston University School of Public Health, Concern for Dying, and the Rubin Family Fund made this edition possible.

Introduction

The most powerful concept shaping the practice of modern medicine is the recognition that patients have human rights. Respect for these rights can transform the doctor–patient relationship from one characterized by authoritarianism to a partnership and simultaneously improve the quality of medical care. The increasing sophistication of medical technology has tended to distance the doctor from the patient and to make the hospital an alien and alienating environment. By demanding that patients be treated as unique human beings, the recognition of human rights in health care can humanize both the hospital and the encounters with physicians and other health care professionals.

Although we try not to think about it, medical care is central to many major life events: from birth to death, from accidents and disability to illness and incapacitation. In the United States, thirty-five million Americans are hospitalized annually, and we spend more than one-half trillion dollars on medical care each year, 12 percent of our gross national product. We worry about our lives, our health, and our money. We often consider our rights only after something has gone wrong; and then lawsuits are high on our list of possible reactions. This is a mistake. Preventing violations of our rights is a much more constructive and effective approach than legal "treatment."

We must all be active participants in making medical treatment decisions, because these decisions can have such a profound effect on our bodies and our lives. Medical decisions are fundamentally personal decisions; different people place different values on longevity, risk, functioning, and bodily appearance. Economics and technology often seem to dictate choices—and the recognition of human rights is the only force powerful enough to prevent medicine from becoming an impersonal and dehumanizing industry.

This book is built on three fundamental premises: (1) the American medical consumer has rights that are not automatically forfeited on entering a health care facility or a doctor–patient relationship; (2) many facilities and physicians fail to recognize the existence of these interests and rights, fail to provide for their protection and assertion, and limit their exercise without effective recourse to the patient; and (3) a doctor–patient partnership in which patient rights and personhood are respected is the most beneficial model for medical decision making for both patients and physicians. Patients are not required to check their rights, along with their other valuables, when they enter a health care institution. Human rights merit strong protection by their very nature. And the protection of human rights in health care can improve the quality of life of both providers and patients, and the quality of care itself.

Many of the legal rights of patients have been enunciated by the courts, often in the context of malpractice suits brought against physicians and hospitals by injured patients. Successful suits demonstrate that the physician not only injured the patient, but did so as a result of negligent conduct. Cataloging these cases, as is necessary to expose and explain the legal rights of patients, can present health care providers and institutions in a less-than-favorable light. Knowledge of these cases, however, is essential to an understanding of the rights patients have, and why they have been enunciated and supported by the courts. It is not the purpose of this book to castigate medical professionals for past misdeeds, it is as inaccurate to term the support of patient rights "antimedicine," as it is to term support of the Bill of Rights "antiAmerican." Nor is being pro-patient in any way "antidoctor." All doctors worthy of the title are pro-patient, and those who are not deserve no respect. The purpose of this book is to inform both patients and their providers of their legal rights in the hope that this information will encourage a more open and mutually satisfying physician–patient partnership.

The first edition of this book, published in 1975, was entitled *The Rights of Hospital Patients*. This edition deletes the adjective. Hospital care accounts for significantly more than half of all health care expenditures (when physician charges for hospital care are added to hospital charges), and most serious injuries to patients occur in hospitals. But there is more to medicine than hospital care; and the basic rights people have are based not on *where* they are treated but on *what* they are. Patients have rights because they are persons, not because they are in or out of a hospital. Although the focus of the hospital is retained in this book, as it is in medicine, the rights discussed also apply in the physician's office, outpatient clinic, HMO surgicenter, nursing home, and other health care settings as well.

The book has been entirely revised and updated to reflect the medical and legal developments of the past decade and a half. The chapter titles are descriptive of the subject matter covered in each. They have been completely rewritten, but the format of the first edition is followed with only two exceptions. The first is that the chapter on payment has been eliminated. Obviously, payment is a critical issue, but it is strictly speaking an "economic rights" issue rather than a "legal rights" issue. The "right to health care" should be recognized as an economic entitlement in this country, and it is a national disgrace that 40 million Americans have no health insurance, and millions of others have inadequate coverage. The AIDS epidemic highlights the urgency of addressing this issue by showing how unfair and inefficient our current system is. Nonetheless, the "right to health care" is distinct from "rights in health care"; and the latter is what this book is all about. The second is that separate chapters on women and children have been eliminated. The subjects previously covered in these chapters now appear in the chapters "Pregnancy and Birth" and "Surgery." Issues specific to mental health are dealt with in another book in this series.

Most of the source material cited in the notes is available in law and medical libraries. Appendix E contains a guide to these libraries for those interested in locating original source material or conducting their own research of a topic. Appendix A lists other resources consumers can turn to, including organizations involved in patient rights and an annotated bibliography. Appendixes B and D set forth "model" legislation on patient rights in general and the right to refuse treatment in particular, and Appendix C contains a sample "living will."

   This book is primarily a reference book for patients, their friends and family, and health care providers. It has also been used as a textbook for medical students and nursing students and should be required reading for medical house staff. Even if you are unable (or unwilling) to read the book, people have reported that their care in hospitals has improved simply by having the book on their bedside table. There is no controlled study on this, but rights safeguard us all, sometimes even when we are unaware of their existence. This book is written to encourage patients to exercise their rights and to help health care professionals recognize and respect these rights. The goal is to resist the dehumanizing effects of medical technology and cost-containment by reasserting the primacy of the individual patient. As was the first edition, this book is again dedicated "to the sick, and the healers who treat them as persons."

1
The Patient Rights Movement

The patient rights movement is as slow as a glacier, equally relentless at changing the landscape, but ultimately healthy. That it is not as organized and identifiable as other consumer movements is explained by the fact that when individuals are sick or injured, they are not themselves, and their first priority is to regain their health and their identity. F. Scott Fitzgerald perhaps put it best in his masterpiece of American fiction, *The Great Gatsby:* "[There is] no difference between men, in intelligence or race, so profound as the difference between the sick and the well." Thus, sick and injured citizens often voluntarily relinquish rights they would otherwise vigorously assert in the hope that this will aid their recovery. In the words of Dr. Oliver Wendell Holmes, "There is nothing men will not do, there is nothing they have not done, to recover their health and save their lives."[1] He describes some of the extremes to which sick people have gone, including being half drowned, half cooked, seared with hot irons, and crimped with knives.

There are additional reasons why there are few organized groups of patient–consumers that apply specifically to hospitalized patients. First, most people in a hospital are sick and are not physically or psychologically capable of either exercising their own rights or organizing to help protect and assert the rights of others. Second, the average length of stay in most American acute-care hospitals is less than a week, hardly long enough to build any type of an inpatient organization. Third, when outside the hospital, most Americans prefer not to think about hospitalization, just as we prefer not to think about death, and so do not plan for it.

Local patient rights organizations have been built around some neighborhood health centers, and by individuals interested in specific populations (such as children), specific occurrences (such as childbirth), or specific diseases (such as kidney failure or AIDS). There is even one national patient rights organization, the People's Medical Society. On the whole, however, the movement for rights in health care is diffuse and unstructured.

The attention of the American people, however, is centering on patient rights primarily because of the increases in the power of medicine's ability to cure disease and delay death. Other important concerns are access to medical care, its quality, and its rising cost. Access concerns have prompted greater enthusiasm for some form of comprehensive national health insurance. Medical advances in the areas of life-prolonging technology, prenatal diagnosis, organ transplantation and artificial implants, and the increased specialization in medicine have all tended to increase the technological and decrease the human aspects of medical care. To maintain a balance in which decision making is shared and the patient retains the right and responsibility to make the ultimate decisions regarding personal care and medical treatment, explicit recognition and protection of the legal rights of patients are essential.

Why are rights especially important in the hospital setting?

Rights are important everywhere. Rights protect personhood and the individual's own value system. Rights help prevent people from being treated as interchangeable, inanimate objects and insist that they be treated as unique persons. But rights have special importance in hospitals, as they do in other "total institutional" settings.

Although no social stigmatization attaches to admission, patients almost never come to the hospital voluntarily. Some uncontrollable or unexpected event, usually an illness or injury, has made admission mandatory. Unless the patient enters through the emergency room, a doctor is likely to have ordered the admission and chosen the particular hospital. Upon arriving, the patient is made to sign a variety of forms that are generally explained only with the assurance that they are "routine." The patient is then separated from accompanying friends or relatives and escorted to a preassigned room. The patient's clothes are replaced with a johnny, a one-piece garment designed for the convenience of hospital staff to make testing and treating the patient easier. The patient is given a plastic wristband with a number written on it, a number that may become more important than the patient's name. The patient is confined to a bed and may even have to await permission to use the toilet.

Medication and food may be prescribed without consulting the patient. Nurses, students, aides, and physicians may enter the patient's room without knocking and submit the patient to all manner of examination and treatment without explanation. Moreover, all this is carefully recorded in a written record that the patient is not generally allowed to see, but which is available to almost anyone on the hospital staff and may be seen as well by medical researchers and insurance companies. Unless the patient is in a private room, visiting hours are usually restricted.[2]

The hospital is an alien environment to most people, and the hospital experience tends to intimidate and disorient and to discourage any assertion of individual rights. Medical care in the past was usually a one-to-one relationship involving only the patient and physician. In the modern hospital, however, the patient is treated by a team of physicians, nurses, and technicians in a complex, unfamiliar, and sometimes frightening setting. Because of this, the institution-patient relationship has become at least as important as the doctor-patient relationship. A firm commitment to individual rights may prevent the dehumanizing influence of the institutional experience from becoming its dominant characteristic.

Why should anyone care about rights in the health care setting?

The answer to this question may seem self-evident, but many health care providers believe that patients should concentrate on getting better and not worry about asserting their rights. And doctors should attend to the patient's welfare: In a commonly heard phrase, "Patients have needs, not rights." In fact, they have both. "Health" and "life" are the primary values espoused by health care professionals. But the right to make one's own choice about treatment options, sometimes called "autonomy" or "self-determination," is even more important to most people. Nor are these views of health providers and their patients uniquely American.

Literature from around the world informs us that many physicians cannot understand why patients seem to take personal decision making so seriously, when the physicians *know* that fighting death and teaching others to fight death are so much more important. In Aleksandr Solzhenitsyn's *Cancer Ward*, for example, a patient, Oleg Kostoglotov, discovers he has been receiving hormone therapy for cancer without his knowledge: "By some right—it doesn't occur to them to question the right—they are deciding for me, without me, on a terrible treatment, hormone therapy. This is like a red-hot iron which, once it touches you, leaves you maimed for life. *But it appears so ordinary in the ordinary life of the hospital!*" (emphasis added).

Oleg fully understands that his physicians believe the treatment is absolutely necessary if he is to survive, but he also understands much more fully than they that there are fates worse than death: "What, after all, is the highest price one should pay for life? *How much should one pay, how much is too much?*...—to save one's life at the cost of surrendering everything that gives it color, flavor, and sparkle. To get a life of digestion, breathing, muscular, and mental activity, and nothing more? To become a walking husk of a man—isn't that an exorbitant price?" (emphasis added).

Most Americans believe they should be involved in medical treatment decisions and at least retain the right to refuse *any* medical intervention. Nonetheless, most of us are concerned that our rights may not be honored in the hospital. Larry McMurtry's description of Emma's feelings on being hospitalized for cancer in *Terms of Endearment* conveys an almost universal resignation: "From that day, that moment almost, she felt her life pass from her own hands and the erring but personal hands of those who loved her into the hands of strangers—and not even doctors, really, but technicians: nurses, attendants, laboratories, chemicals, machines."

This is powerful imagery, powerful primarily because of its accuracy. But it does not have to be like this. We do not have to choose between expert technicians and humane caretakers. We can expect better when we deliver ourselves "into the hands of strangers," and the articulation and assertion of patient rights can help make a humanized hospital experience possible.[3]

But if we want to be treated like individuals in hospitals, we will have to demand humane treatment. Even physicians dedicated to humanizing the hospital experience tend to blame the public, not the medical profession, for the current state of affairs. Melvin Konner's argument is typical:

Most Americans are, so far, unwilling to sacrifice scientific and technical perfection even in these limited realms, believing, perhaps mistakenly, they will fare better by sticking it out in the great hospital-palaces, however cold and forbidding. For the treatment of most illnesses, for which technical knowledge and prowess make a difference, we seem to prefer a cold or even a disturbed physician with full command of current medical sciences, to the most sensitive and compassionate bumbler. Psychologically, we seem to tolerate anything, so long as the doctors make the pain go away, so long as we leave the hospital alive.[4]

But this dichotomy is a false one—there should be no need for anyone to have to choose between a "cold" and "disturbed" technician and a "sensitive and compassionate bumbler." If these are the only two types of physicians current medical education can produce, radical change is required. And if doctors really believe patients like to be treated like inanimate objects or pets, then we all have a responsibility to tell them firmly and unequivocally that they are mistaken: We want and deserve physicians who are both knowledgeable *and* compassionate.

Does it make any sense to talk about rights in health care?

It does. Even though there are many other approaches that merit attention, the concept of using patient rights to make service to the individual patient the core of health care is crucial to any humane and responsive health care system. Another way to change the role of the patient in health care delivery is to concentrate on wider questions, such as our economic and governmental system. Some critics believe that fundamental change in the provider–patient relationship (to make it more responsive and equitable) is possible only after there is a basic change in our social structure. The argument is that the health system is primarily a mirror of the larger social, political, and economic system of which it is a part, and that changes in health care policy and practices will come only from changes in that larger system. Problems of equity and access are thus viewed not as problems of physician discrimination or hospital policy but as problems of poverty, social class, race, and geographic location.

There is much merit in this argument, and not every problem can be entirely resolved within the provider–patient relationship. Providers not only have formal relationships with their patients but also have relationships with other providers, health care institutions, and numerous governmental agencies. A provider's relationship with these institutions and individuals is often a very complex one, and providers themselves are often confused and therefore submissive in cases where they do not understand their own rights. As health care has become the major service industry in the United States and government has become one of the major sources of funding for the industry, regulatory agencies and review committees have become increasingly significant.

Whether an operation is covered by insurance or by a health maintenance organization (HMO), may depend on an interpretation by a clerk or on the review of its "necessity" by other health care professionals; whether medical research may be done may depend on a determination of the hospital's Institutional Review Board (IRB) and the federal government's Food and Drug Administration (FDA); whether a medical student may practice in a certain setting may depend on state statutes and licensing regulations; and whether husbands are permitted in the delivery room may depend on hospital policy.

In all these cases, both the health care practitioners *and* the patient will be better off if the status of the law regarding both patient *and* provider rights is understood, and the means of change or challenge well delineated. I would go even further. Because it informs physicians of such things as when they have the authority to treat minors and emergency patients, and when a decision made by an informed patient can be acted on without fear of criminal or civil liability, an understanding of the law and legal rights can be as important to the proper care of patients as an understanding of emergency medical procedures or proper drug dosages.[5]

Whether our social, political, and economic system is changed in the future or not, we must all live in the real world today. In this world, not only talking about rights makes sense, but asserting rights is often the *only way* to accomplish specific objectives.

What is meant by having a "right" to something?

A right to something is a *valid claim* to it.[6] If you have bought vitamins, you can make a claim to their use; you are *entitled* to use them, to mash them up and eat them, for example, or to throw them away. Your claim to the vitamins is valid because it is backed by law. If there were no laws concerning property rights, your claim might be made valid by appeal to some secular or religious moral rules. Some laws are unequivocal and certain, having been interpreted and applied by courts. Others are less clear because they have not been specifically declared by the courts or legislatures. These rights have "probable" legal status, if it is probable that they would be upheld if tested in court.

Three kinds of rights are discussed in this book: *legal rights,* or claims that would be currently backed by law if the case went to court; *probable legal rights,* which would likely be backed by law if the case went to court; and *human rights,* which are critical to maintaining human dignity but have not yet attained legal recognition.

As these examples demonstrate, there is no single or absolute definition. To understand any definition, one must know the purpose for which the definition is sought, the audience for which it is intended, and the identity of the definer. In regard to the concept of a right, it is most helpful to think of a continuum. At one end would be all of those rights that are recognized as legal rights. These include the *rights of citizenship* arising under the Constitution and its Amendments, the laws of the fifty states, and court decisions. One legal scholar referred to such a right as "a capacity residing in one man of controlling, with the assent and assistance of the state, the actions of others."[7] This is a *legal right.*

Somewhere near the middle of the continuum are those rights that, with a high degree of probability, would be recognized as legal rights by a court of law. In most situations, all that is needed is the appropriate case to present the court with the opportunity to recognize a new legal right. This type of right involves a reasonable expectation of what a court would do if called on to deal with the issue. This is a *probable legal right.* It approximates Oliver Wendell Holmes' predictive theory of the law: "The prophecies of what the courts will do in fact, and nothing more pretentious, is what I mean by law." [8]

At the other end of the continuum are statements of what the law ought to be, based on a political or philosophical conception of the nature and needs of humans. In making a declaration of what we believe should be, we are making a political statement. Such rights may be considered of fundamental importance and preexist recognition by positive law. The civil rights movement in the United States provides numerous examples, as does the United Nations Universal Declaration of Human Rights. This may be termed a *human right*.

This book is primarily concerned with legal rights and probable legal rights. But at times, the formal recognition of human rights (those not currently recognized by law) will be advocated when their existence is important to patients.

Rights give us dignity and protection. If we have a right to something, we can insist on it without embarrassment or fear: We do not, after all, feel embarrassed in demanding our right to vote. Nor do we depend on our neighbor's kindness in not peeking through our window or going through our mail. Patients need not request, but should insist, that they be treated as individuals.

Nevertheless, it may not be sensible to demand rights under all circumstances. Patients may (unfortunately) run the risk of irritating the medical and nursing staff if they insist too much upon exercising all of their rights, but patients may find it necessary to demand their rights nonetheless. If patients exercised their rights more often, such insistence would no longer be seen as unusual, and it would become increasingly easy for other patients to exercise their rights as well.

What rights do patients have?

The following "model" Patient Bill of Rights lists the primary rights that should be accorded all patients both as policy by health care facilities and as law by state legislatures. The term "rights" is used in the three senses described in the answer to the preceding question. Where the phrase "legal right" is used, the reference is to a legal right. The term "right" refers to a probable legal right, and "we recognize the right" refers to a human right. Once these rights are recognized, some mechanism for handling complaints and enforcing rules must also be established. If enacted into law or accepted as facility policy, all of these rights would then become legal rights.

The model bill is set out as it would apply to a patient's chronological relationship with a hospital or other inpatient facility: sections 1–4 for a person not hospitalized but a potential patient; 5 for emergency admission; 6–17 for inpatients; 18–24 for discharge and after discharge; and 25 relating back to all twenty-four rights. The specific rights mentioned are all discussed elsewhere in this book.

A Model Patient Bill of Rights

Preamble: As you enter this health care facility, it is our duty to remind you that your health care is a cooperative effort between you as a patient and the doctors and hospital staff. During your stay a patient rights advocate will be available to you. The duty of the advocate is to assist you in all the decisions you must make and in all situations in which your health and welfare are at stake. The advocate's first responsibility is to help you understand the role of all who will be working with you, and to help you understand what your rights as a patient are. Your advocate can be reached twenty-four hours a day by dialing_____. The following is a list of your rights as a patient. Your advocate's duty is to see to it that you are afforded these rights. You should call your advocate whenever you have any questions or concerns about any of these rights.

1. The patient has a legal right to informed participation in all decisions involving the patient's health care program.
2. We recognize the right of all potential patients to know what research and experimental protocols are being used in our facility and what alternatives are available in the community.
3. The patient has a legal right to privacy regarding the source of payment for treatment and care. This right includes access to reasonable medical care without regard to the source of payment for that treatment and care.
4. We recognize the right of a potential patient to complete and accurate information concerning medical care and procedures at our facility.

5. The patient has a legal right to prompt attention in an emergency situation.
6. The patient has a legal right to a clear, concise explanation in layperson's terms of all proposed procedures, including the possibilities of any risk of mortality or serious side effects, problems related to recuperation, and probability of success, and will not be subjected to any procedure without the patient's voluntary, competent, and understanding consent. The specifics of such consent shall be set out in a written consent form and signed by the patient before the procedure is done.
7. The patient has a legal right to a clear, complete, and accurate evaluation of the patient's condition and prognosis without treatment before being asked to consent to any test or procedure.
8. The patient has a right to designate another person to make health care and treatment decisions for the patient, and based on the patient's own directions and values, in the event the patient is unable to participate in decision making. The health care facility agrees to recognize the authority of an individual so designated.
9. The patient has a right to know the identity, professional status, and experience of all those providing service. All personnel have been instructed to introduce themselves, state their status, and explain their role in the health care of the patient. Part of this right is the right of the patient to know the identity of the physician responsible for the patient's care.
10. The patient has a legal right not to be discriminated against in the provision of medical and nursing services on the basis of race, religion, national origin, sex, or handicap.
11. Any patient who does not speak English or who is hearing impaired has a right to have access to an interpreter.
12. The patient has a right to all the information contained in the patient's medical record while in the health care facility, and to examine the record on request.

13. The patient has a right to discuss the patient's condition with a consultant specialist, at the patient's request and expense.
14. The patient has a legal right not to have any test or procedure, designed for educational purposes rather than the patient's direct personal benefit, performed on the patient.
15. The patient has a legal right to refuse any particular drug, test, procedure, or treatment.
16. The patient has a legal right to privacy of both person and information with respect to: the staff, other doctors, residents, interns and medical students, researchers, nurses, other health care facility personnel, and other patients.
17. We recognize the patient's right of access to people outside the health care facility by means of visitors and the telephone. Parents may stay with their children, and relatives with terminally ill patients, twenty-four hours a day.
18. The patient has a legal right to leave the health care facility regardless of the patient's physical condition or financial status, although the patient may be requested to sign a release stating that the patient is leaving against the medical judgment of patient's doctor or the staff.
19. The patient has a right not to be transferred to another facility unless the patient had received a complete explanation of the desirability and need for the transfer, the other facility has accepted the patient for transfer, and the patient has agreed to the transfer. If the patient does not agree to the transfer, the patient has the right to a consultant's opinion of the desirability and necessity of the transfer.
20. A patient has a right to be notified of impending discharge at least one day before it is accomplished, to a consultation by an expert on the desirability and necessity of discharge, and to have a person of the patient's choice notified in advance.
21. The patient has a right, regardless of the source of payment, to examine and receive an itemized and detailed explanation of the total bill for services rendered in the health care facility.

22. We recognize the right of a patient to competent counseling from the facility staff to help in obtaining financial assistance from public or private sources to meet the expense of services received in the health care facility.
23. The patient has a right to timely prior notice of the termination of eligibility for reimbursement by any third-party payer for the expense of care.
24. At the termination of the patient's stay at the health care facility the patient has a right to a complete copy of the information contained in the patient's medical record.
25. We recognize the right of all patients to have twenty-four-hour-a-day access to a patient rights advocate who may act on behalf of the patient to assert or protect the rights set out in this document.

Can sick and injured patients really exercise their rights?

As is apparent from the preamble of this document, a statement of rights is insufficient. What is needed in addition is someone, who may be termed an advocate, to assist patients in asserting their rights. Parents are the natural advocates for their children. One's spouse may be seen as a natural advocate; and one's children, in the case of the elderly, may also be their natural advocates. Family relationship alone, however, should not determine advocate status. The patient, if able, should be the one to designate an advocate, who may be a best friend, personal physician, attorney, or other person he or she has trust and confidence in. The main point is that when a person is sick or injured, the person is not in a realistic position to assert rights and thus runs the risk of being treated as a nonperson. Doctors, nurses, and social workers all like to think of themselves as the primary advocate for their patients, and many actually perform this role well. But only individual patients can decide whom they trust enough to help them exercise their rights. Since many people have no one who can act effectively as their advocate, health care facilities should employ individuals capable of fulfilling this role for patients. The concept of the advocate is dealt with in detail in chapter XV.

Is there a legal right to health care in the United States?

No. Although specifically recognized in proclamations of both the United Nations and the World Health Organization, the right to health care appears neither in the US Constitution nor in our Declaration of Independence (although "life, liberty and the pursuit of happiness" are all made rather difficult without proper health care). The strongest congressional expression on the subject is found in the preamble to the 1966 Comprehensive Health Planning Act, which states that "the fulfillment of our national purpose depends on promoting and assuring the highest level of health attainable for every person." Persons who qualify for Medicare and Medicaid enjoy a statutory right to have some of their medical bills paid, but there is no general legal right to demand medical services (except in emergency situations). The adoption of a program of national health insurance would, by itself, serve only to expand the payment right without addressing itself to the problem of access to services. Nonetheless, since access is often determined by ability to pay, a comprehensive program of national health insurance would certainly be helpful to the almost forty million Americans (most under the age of twenty-four) who currently have *no* health insurance.[9]

What characteristics of the present doctor–patient relationship make it difficult for patients to assert their rights?

There are at least four characteristics of the doctor–patient relationship in decision making that make it difficult for patients to retain the power to make the final decisions concerning their care:

1. Ambiguity that a decision must be made and that options exist
2. Ambiguous identification of the decision maker
3. Ambiguous identification of the person or entity that commands the decision maker's loyalty
4. Control of pertinent medical information by the physician

This list is neither all-inclusive nor universal; nevertheless, patients can, by using it as a checklist, determine just how likely it is that they will be able to make the final decision concerning what treatment, if any, they will be given. For example, the patient may not realize that options (such as the right to refuse treatment) or alternatives (such as less radical surgery) exist (characteristic 1). The real decision maker may be a patient's spouse or family (characteristic 2); the physician's primary loyalty may be to the parents or (much more unusually) to a colleague engaged in a research protocol rather than a child–patient (characteristic 3); and the physician may be reluctant or unwilling to share information regarding diagnosis or prognosis with a patient (characteristic 4). To help improve the chances of getting the care they want, patients may have to insist that they themselves make all important decisions, that their family be consulted only with their permission, and that they personally review all of their medical records.[10]

Can patients ever regain a major role in decision making?

Shared decision making with health care providers is a realistic goal and one that should be striven for. In addition, the patient must always retain the right to refuse any specific recommended treatment. The objection that the legal system has no legitimate interest in "interfering" with the current doctor–patient relationship is meritless. Even a cursory glance at the history of the past century demonstrates the inherent weakness of such a position.

There are an extraordinary number of precedents demonstrating that the law has been encouraging a shift of power in relationships that have traditionally been protected by the law, in some cases, for centuries. Examples include buyer–seller, landlord–tenant, debtor–creditor, employer–employee, warden–prisoner, police–suspect, teacher–student, parent–child, and those involving minorities, such as blacks, children, and the elderly.

Several things are evident from this listing. First, the mere existence of a relationship, established by force of law or by force of habit, is not itself sufficient justification for its continuation. Second, when there is sufficient justification, in terms of basic constitutional rights or fundamental human fairness, the legislature, the judiciary, and the executive can and will act to redefine even traditionally protected relationships. Third, the process of redefinition may take many paths. It may involve a statute, injunctive relief, or an executive order. No matter the form, however, the effect is qualitatively the same. [11]

Legal redefinition of the doctor–patient relationship is neither a radical nor unprecedented suggestion. A movement away from the doctor simply doing what he or she thinks is best for the patient (sometimes called "paternalism") and toward shared decision making with the patient is already well under way in this country. Prior to the 1970s, medical paternalism was the rule. The 1970s and early 1980s witnessed the blossoming of the doctrine of informed consent, with its requirement that certain information be shared with patients. We are only now beginning to take information sharing and decision sharing seriously, and it will take some time before it becomes routine. But the direction is set, and the articulation of patient rights prods us all more quickly down the path toward true shared decision making in health care and medical treatment.[12]

NOTES

1. O. W. Holmes, "The Young Practitioner," in W. H. Davenport, ed., *The Good Physician* (New York: Macmillan, 1962), at 176. See also E. J. Cassell, *The Healer's Art* (Cambridge, Mass.: MIT Press, 1985), at 25–6.
2. Annas, *The Hospital: A Human Rights Wasteland*, 1(4) Civil Liberties Rev. 9 (1974). See also C. B. Inlander, L. S. Levin & E. Weiner, *Medicine on Trial* (New York: Prentice-Hall, 1988).
3. G. J. Annas, *Judging Medicine* (Clifton, N.J.: Humana Press, 1988), at 2-3.
4. M. Konner, *Becoming a Doctor* (New York: Penguin, 1988), at 371.
5. G. J. Annas, L. H. Glantz & B. F. Katz, *The Rights of Doctors, Nurses and Allied Health Professionals* (Cambridge, Mass: Ballinger, 1981), at xiv-xvi.
6. J. Feinberg, *Social Philosophy* (New York: Prentice-Hall, 1973).
7. T. E. Holland, *Elements of Jurisprudence*, 12th ed. (London: Oxford U. Press, 1916).
8. Holmes, *The Path of the Law*, 10 Harv. L. Rev. 457, 461(1897); and see L. L. Fuller, *The Morality of Law* (New Haven, Conn.: Yale U. Press, 1964), at 106-7.
9. See e.g., R. Fein, *Medical Care, Medical Costs: The Search for a Health Insurance Policy* (Boston, Mass.: Harvard U. Press, 1986); Himmelstein &

Woolhandler, *A National Health Program for the United States: A Physicians' Proposal*, 320 New Eng. J. Med. 102 (1989); Enthoven & Kronick, *A Consumer-Choice Health Plan for the 1990s*, 320 New Eng. J. Med. 29 (1989); Relman, *Universal Health Insurance: Its Time Has Come*, 320 New Eng. J. Med. 117 (1989); and National Leadership Commission on Health Care, *For the Health of the Nation* (Washington, DC: NLCHC, 1989).

10. These characteristics are discussed in more detail in Annas, *Medical Remedies and Human Rights: Why Civil Rights Lawyers Must Become Involved in Medical Decision-Making*, 2 Human Rts. 151, 156–64 (1972).

11. *See* Annas & Healey, *The Patient Rights Advocate: Redefining the Doctor-Patient Relationship in the Hospital Context*, 27 Vand. L. Rev. 243 (1974). For critical comment, see Margolis, *Conceptual Aspects of a Patients' Bill of Rights*, 11 J. Value Inquiry 126 (1977). For a summary of competing theories of law, *see* M. Kelman, *A Guide to Critical Legal Studies* (Cambridge, Mass.: Harvard U. Press, 1987).

12. *See* ch. VI, "Informed Consent," for a detailed discussion of this trend. See also Leenen, *Patients' Rights in Europe*, 8 Health Policy 33 (1987); and President's Commission for the Study of Ethical Problems in Medicine and Biomedical and Behavioral Research, *Making Health Care Decisions* (Washington, DC: Government Printing Office, 1982).

II
How Hospitals Are Organized

Describing how hospitals are organized and administered is somewhat like describing high schools. The differences in their size, location, and special programs may be more important than their similarities in organizational structure. Nevertheless, an introduction to hospitals is useful as long as it is remembered that the generalities must be modified by the idiosyncracies of each hospital or other health care facility.

What kinds of hospitals are there?

The approximately 6800 hospitals in the United States can be classified either by type of ownership or service. About 2000 are owned by the local, state, or federal government. Of the remainder, about 1000 are for-profit, private hospitals, also termed *proprietary hospitals*. Many of these are a part of a "chain" of hospitals with one corporate owner. The majority of other hospitals are private, nonprofit institutions, also called charitable or voluntary hospitals. These hospitals have approximately half of the approximately 4 million hospital beds and account for more than two-thirds of the approximately 35 million annual hospital admissions and more than half of the approximately 300 million annual outpatient visits.[1]

Hospitals may specialize in particular services, such as pediatrics or obstetrics, or in diseases such as cancer, but the trend is to be "general," capable of handling most medical problems. If a hospital is affiliated with a medical school and has a teaching program for medical students, interns, and residents, it is also a "teaching hospital."

Who runs the hospital?

Hospitals are corporations and thus follow the corporate model of organization. Whatever management model is used, a board of trustees or a board of directors is in charge and provides overall guidance and direction to the hospital administration. Depending on the particular hospital, the involvement of the board in the daily operations can vary from detachment to requiring approval of all but routine matters. It is not unusual for a board to be composed of local

17

community leaders, such as bankers and lawyers, who are passive and have no direct involvement in the institution. The board hires and relies on the members of the hospital's administration to meet the institution's goals and objectives. The chief administrator is usually referred to as the director or the president. This individual generally has a background in health care administration and often possesses an MBA (Master of Business Administration) or MPH (Master of Public Health). Occasionally, the director is an MD (Doctor of Medicine). The director usually has a number of assistant and associate directors, each with a particular area of authority. When all else fails, patients may need to seek the direct intervention of the hospital administrator. There will be an administrator in the hospital or "on call" twenty-four hours a day. The administration is responsible for the day-to-day operation of the hospital.

The delivery of medical care is the responsibility of the medical staff, which in its own turn answers directly to the board of trustees or directors. A joint committee, with representation of both administration and the medical staff, often provides a liaison between the two groups. The medical staff has a great deal of authority and autonomy as the group with medical knowledge and the group responsible for generating revenue by filling the hospital's beds with patients. Nonetheless, the board retains the ultimate legal authority and responsibility regarding hospital policy.

How are the hospital's goals determined?

For decades, economists assumed that major corporations were primarily guided by profits. John Kenneth Galbraith has made a rather persuasive argument that once enough profits have been generated to assure survival, large firms have other goals, usually to maximize growth "consistent with the provision of revenues for the requisite investment." Moreover, as an industry grows more complex, defining the nature of these goals and how they will be implemented is no longer in management's hands. Rather control rests with production and design experts, whom Galbraith terms the "technostructure." [2]

An analogous movement can be seen in many large hospitals, both profit and nonprofit. Once they have reached the point at which they can deliver a wide range of competent health care services and have been accepted by the community as quality hospitals, their goals tend to shift and multiply. One common goal is the development of expertise in specialized areas, often backed by increasingly sophisticated and expensive medical equipment. Like the corporate technostructure, the senior staff physicians are often able to dictate to hospital management the types of equipment and staffing necessary to develop these specialized capabilities, such as heart transplantation capability and sophisticated imaging equipment. Such glitzy technology adds to the prestige of both the institution and the members of its medical staff.

This institutional quest for glory, with its demands of specialization and subspecialization, makes sense on a regional or multistate level. When every hospital in a community, however, acquires the same array of expensive and redundant equipment, waste of medical resources is inevitable. But the progression is inexorable after the hospital decides to add to its goal of maximizing the health of the community, the goal of internal growth of specialized expertise.

Others have defined the multiple goals of the hospital as follows:

- Caring for the sick and injured by diagnosing and treating their ailments and relieving their pain and suffering
- Providing an educational facility for doctors, nurses, and other health care personnel
- Promoting public health in the community by preventing disease and accidents
- Encouraging active research in the field of human medicine

As will become apparent in later chapters, these goals are not always completely compatible, and patient care may be compromised by those who rank prestige, education, or research ahead of the treatment of individual patients. Nonetheless, as the former director of the Massachusetts General Hospital has emphasized, "The first function of any hospital worthy of the name is to serve the sick and injured." [3]

Who is in charge of an individual patient's care?

If a patient has a personal physician who is able to admit patients to the hospital (i.e., one who has "privileges" at the hospital), then the patient's personal physician will likely be the patient's attending physician. The attending physician retains ultimate authority and accountability for managing the patient's care. If the patient does not have an attending physician, the patient will be in a category called "service" patients and care will be managed by a resident physician of the appropriate service, such as medicine, surgery, or obstetrics. Problems can arise when patients admitted through a clinic or an emergency department do not know who is in charge of their care. The hospital has a duty to put someone in charge of the patient's care, and the patient has a right to know who that person is.

It is easy to understand the importance of choosing a physician ahead of time so that in an emergency a person will feel comfortable with the physician who is managing the care. It is, of course, wise to choose a physician who has admitting privileges at the hospital in which a person wants to be treated.

What is a teaching hospital?

Teaching hospitals are affiliated with medical schools and provide training sites for their students, and for recent graduates of their school and other medical schools. Physician coverage is provided by house staff who consult with the private or attending physician at least daily and, with increasing frequency, in-house specialists in anesthesiology, surgery, emergency medicine, and radiology. House staff in their first postgraduate year (PGY) of training may be called PGY-1s, interns, or first-year residents. Supervision is provided by residents who are usually at least in their second postgraduate year (PGY-2) of training. In contrast, in the nonteaching community hospital the nursing staff must respond to emergency patient needs alone, at least until they can contact the attending physician.

Twenty-four-hour-a-day physician coverage is an obvious benefit. In addition, new physicians and medical students are steeped in the current literature and most recent advances in medical science. Students frequently have time to listen to patient concerns and to assess and respond to needs. The time and concern they devote to patients can provide greater insight, and it is not unusual for students to uncover problems that are overlooked by others. On the other hand, teaching hospitals can be tiresome and intrusive, since patients are frequently subject to repetitive examinations and interviews. Therefore, it is important to choose a hospital based on what the hospital can provide the patient. An unusual or complex problem generally warrants a teaching setting, where the latest technology and the most up-to-date specialists are available. A simple hernia repair or nonsurgical reduction of a fracture can be adequately managed in the local community hospital.

Who are the "house staff" in a teaching hospital?

There is a military-like "pecking order" among hospital physicians that reflects education, experience, and job responsibilities. All physicians are not the same and do not possess equal experience and ability.

Day-to-day care in a teaching hospital is managed by the interns and residents, collectively referred to as the house staff. A patient's personal or attending physician generally oversees care. A hospital is a teaching facility by virtue of the presence of house staff who literally provide physician coverage to the hospital or "house" (a term that originated when doctors actually lived at the hospital). At the same time that house staff provide physician coverage, they are developing essential clinical skills in the diagnosis and management of patient problems. They also supervise and teach *medical students*, who are on the lowest rung and frequently limited to doing the routine work that the residents prefer not to do. Ironically, the medical student's admitting note will be the most complete and thorough part of the patient's medical record.

PGY-1s will spend more time with the patient than any other physician. These individuals spend many long and grueling hours at the bedside, and because of their burdensome schedules, there have been proposals to limit the number of hours house staff are on duty.[4] New graduates begin work in July, so patients are likely to be seen by a relatively inexperienced PGY-1 if admitted to a teaching hospital during the middle of summer. July is tough on everyone.

Next in line are PGY-2s, or second-year residents. Residents provide a system of checks and balances to ensure safe patient care. They supervise and provide teaching to the PGY-1s in their specialty, such as internal medicine or surgery. If a patient is uncomfortable with any aspect of care, the PGY-2 should be easily accessible for questions and information.

The patient will see the PGY-2 or PGY-3 at least each morning during "rounds," usually conducted between 7:00 and 10:00 AM or between 4:00 and 6:00 PM. At that time a team, usually composed of the PGY-2, several PGY-1s, medical students, and nurses, will go around to visit each patient. The patient may or may not be examined but, at a minimum, should expect to have some communication with the team. It is not unusual for a senior member of the medical staff to participate in rounds. This is an excellent time to ask questions and request information, so patients should be prepared and not hesitate to speak up. Patients should never allow discussion to occur over or around them: Patients have a right to participate in, and to ultimately determine, their own care. It is often useful for patients to write down questions to ask during rounds, although this practice irritates some paternalistic physicians.

A final system of checks and balances is provided by the *chief resident* and a staff physician responsible for house staff training. Patients will often meet these people on rounds and can easily request to speak with them at any time. The chief resident is selected based on experience and demonstrated ability and again provides supervision, direction, and education to house staff as well as managing the schedule. The director of house staff training is responsible for all residents and reports to either the chief or assistant chief of the respective department.

All physicians should wear name tags, and the designation of medical student, PGY-1, PGY-2, etc., should clearly appear. If patients need to contact any of these people, they can ask either house staff, nurses, or the telephone operator for assistance. Some hospitals have a patient representative who can also provide useful information. *Remember that if a patient is dissatisfied, there is always someone else in the physician hierarchy that the patient can talk with.*

Who are the other doctors in the hospital?

While in the hospital, the patient may also be seen by consultants, who are generally subspecialists, who are usually "board certified." Specialization is regulated by the medical profession through

American Specialty Boards. There are about two dozen such boards, each of which requires an applicant to complete an internship and three to seven years of approved training and practice before issuing a certificate, which makes the doctor a "board certified specialist." Although no law requires specialists to be certified, those that have certification are generally the best, since they have specialized training and have agreed to devote the majority of their practice time to their specialty. The specialties include allergy and immunology, anesthesiology, colon and rectal surgery, dermatology, family practice, internal medicine, neurological surgery, nuclear medicine, obstetrics and gynecology, ophthalmology, orthopedic surgery, otolaryngology, pathology, pediatrics, physical medicine and rehabilitation, plastic surgery, preventive medicine, psychiatry and neurology, radiology, surgery, thoracic surgery, and urology. It is the patient's right to know whether the consultant is board certified and exactly why the consultant has been called before the patient consents to be examined.

Interns, residents, and consultants are medical doctors (MDs). Medical students are just that, students training to become doctors. A patient in a teaching hospital, particularly a university hospital, will likely be seen by medical students. They are usually in their third or fourth year of school, but may have just started. They may be introduced as "doctor," "young doctor," or "student doctor." Unless "MD" follows their name on their name tag, they are not physicians. Patients may refuse to be examined or treated by a student or any other health provider. Nonetheless, it is important to recognize that bedside training is a critical part of physician preparation. Patients who can accommodate a request to be examined by a student may be making an important contribution to the student's education.

Patients should also be aware of the "see one, do one, teach one" method of educating medical students and house officers. Patients should not hesitate to ask if medical students or residents have done a proposed procedure (e.g., placement of an I.V. line or a lumbar puncture) and if so, how many times. Patients not wanting to be used for "practice," which can entail both increased pain and risk of injury, have an absolute right to refuse and to have a qualified, experienced person do the procedure instead. Patients and their advocates should not hesitate to learn the qualifications and experience of all those proposing to treat the patient before a procedure is performed.

A medical staff may be divided into a number of classifications, including the following: active staff (regular membership); associate (junior membership); consulting (a doctor who can consult but not admit); courtesy (a doctor who only occasionally uses the hospital); emeritus (retired doctors); house staff (interns and residents); honorary (for noted doctors not heavily involved in the hospital).

Staff appointments are usually renewable annually. In exchange for this privilege, staff physicians are usually required to give some of their time in service to the hospital. This may take the form of committee membership, teaching, research, or covering the emergency ward, and it varies considerably from one hospital to another.

Biographical information on most physicians licensed to practice medicine can be obtained by consulting the *American Medical Directory*, published biennially by the the American Medical Association (AMA). Information can also be obtained from the state or local medical society. For board-certified physicians, *The Directory of Medical Specialists* gives a more comprehensive biography, lists all diplomates of the American Specialty Boards, and gives the requirements for board certification in each specialty. Both of these publications should be available in the hospital library.

Once a patient is in a hospital room, which staff members is the patient most likely to see, and what is the role of the nursing staff?

Patients see a large number of people, particularly in teaching hospitals. The amount of time physicians spend at the bedside increases or decreases depending on the seriousness or acuity of the patient's problem. Generally, only the nursing staff is present around the clock. The department of nursing employs the largest number of staff, and departmental salaries constitute a large portion of the hospital's overall wage and salary program. Nursing and other hospital staff are assigned on scheduled shifts. Most shifts will cover periods of eight to ten hours from 7:00 AM to 3:30 PM, 3:00 PM to 11:30 PM, 11:00 PM to 7:30 AM, or 7:30 AM to 6:00 PM, etc. Recently, various flexible schedules have been made available to staff, including twenty-four-hour weekends (two twelve-hour shifts). Some delay can be anticipated in response time during the change of shift when the floor staff are giving and getting a report on all patients on the unit.

Nurses are licensed professionals, whose primary responsibility is to coordinate the patient's care and to take responsibility for its continuity. They are the only professionals always available twenty-four hours a day (although many large hospitals now have twenty-four-hour-a-day physician coverage as well) and the only ones likely to know about everything that will happen to a particular patient in the hospital. They are often overworked and underpaid, and there is currently a major nurse shortage in many cities and hospitals that will not go away soon. This shortage is related to low pay, low prestige, lack of autonomy, and rotating and weekend shifts.[5] If there is a shortage of nurses in a hospital, both the patient's care and comfort will be compromised, and there is not much the patient or advocate can do about it.

Generally, registered nurses (RNs) assess a patient's needs and plan care to meet these needs. They document both components of this process and evaluate their interventions. Nursing care may include physical care, teaching, and emotional support. Nursing presence at the bedside is vital to close patient observation, monitoring, and continuous care. Many hospitals now provide nursing care using a *primary nursing* model. The so-called primary care nurse is a registered nurse who assumes ultimate or primary responsibility for planning a patient's care from the time of admission through discharge. When the primary nurse is not at work, an associate nurse will carry out the plan of care and communicate all pertinent information about the patient to the primary nurse. This system enhances the overall coordination and continuity of care. Most important, a specific nurse is responsible for the patient's care. If primary nursing is practiced on the patient's floor, the patient should expect to be informed of the identity of the primary nurse and to develop a working relationship with the nurse. *The primary nurse can be the patient's most important advocate.*

If primary nursing has not yet been implemented, then it is likely that a form of nursing called *team* will be practiced. The *team leader* will be an RN who will be responsible for caring for a large group of patients with the assistance of either other RNs, LPNs (licensed practical nurses), aides, or any combination thereof. The team leader will most likely do rounds with the physician team, pass medications, manage intravenous and other therapies, and generally supervise patient care. Patient care is more fragmented in a team-nursing setting. But regardless of the delivery system, there is a chain of command and a channel to use to communicate concerns. On the unit level, there is generally a designated person in charge during each shift. During the evening, at night, on weekends and holidays, this will usually be a staff nurse who has general supervisory responsibilities. This person is called the *charge nurse*. During these times, there will also be a *nursing supervisor* who will provide general supervision to the entire hospital. During the day, the *head nurse* has ultimate accountability for the quality of care delivered. The head nurse may have an assistant. On an administrative level, the head nurse will generally report either to an assistant director of nursing or to the director, depending on the size of the institution.

Many institutions have adopted a "decentralized" approach to nursing administration. This means that authority and accountability for decision making are delegated to the head nurse level. In this type of setting, head nurses interview, hire, counsel, discipline, and even fire employees. Patients who have a concern and are unable to obtain a satisfactory response from the staff nurse should seek out the head nurse or the charge nurse.

Like physicians, nurses are not all the same. Nurses have a variety of backgrounds in education, experience, and responsibility. This fact presents a confusing picture. The basic education preparing nurses to become RNs varies from two to five years. The designation "RN" simply indicates that the individual has completed an educational program and passed a state licensing exam. The RN may have received an associate degree from a two-year college, a diploma from a hospital-based three-year program, a baccalaureate degree from a four-year college, or a master's degree from a graduate school of nursing. Currently, there are several universities that offer doctoral degrees in nursing, although a nurse with a doctorate is unlikely to be seen at the bedside. The education necessary to enter into professional practice is hotly debated among nurses today. The American Nurses Association, nursing's professional organization, has stated that professional practice requires a baccalaureate degree. Many nurses disagree.

The letters "GN" next to a nurse's name indicate a status of graduate nurse and mean that the nurse has not yet passed a state licensing exam. After taking the examination, the GN is often allowed to function within the scope of an RN until notification of a passing grade. In the event of failure, the GN can no longer function as an RN but may be employed as an aide. The letters GN should alert patients that the nurse is a new graduate with limited experience.

Hospitals also have other specialized personnel in such fields as respiratory therapy and physical therapy, transportation, and dietary, who may also see the patient often, depending on the problem. Some hospitals have more than eighty-five separate job descriptions for various medical personnel. Social workers and chaplains are also available for patients who wish to see them. They can be very helpful in resolving specific problems.

How is the hospital's medical staff organized?

The organization of the hospital's medical staff is determined by the medical staff bylaws. These are written rules, adopted by the medical staff itself, that govern the conduct of the physicians who practice in the hospital. Bylaws vary from hospital to hospital, but usually include a statement of purpose, a list of required qualifications of staff members, a descriptive outline of the staff organization, and procedures for appointing, renewing, or removing staff members, a statement concerning staff functions (which will usually be assigned to particular committees), a statement on the types of conditions each class of doctor may care for, and the types of treatments the doctor may use. Finally, there will be a list of staff rules and regulations concerning such things as record keeping, tissue removal and examination, tests that must be performed on all patients admitted to the hospital, consent procedures, and the signing of physician orders.

Who is in charge of the hospital staff?

Usually the members of the medical staff are accountable to the president or chief of staff for the delivery of care. The chief of staff has administrative responsibility for clinical services and often reports to a hospital administrator. The chief is in a powerful position and acts as a liaison among the medical staff, hospital administration, and board of trustees. The position is one of respect and authority.

The staff is often subdivided into specific departments or services such as medicine, surgery, and pediatrics, each with its own chief of service. *If the patient cannot get satisfaction concerning a medical complaint from the attending physician, the chief of service or chief of staff should be able to take appropriate action.* The patient and advocate should not hesitate to call the chief of service or chief of staff if unable to resolve a concern with the attending physician.

What committees in the hospital structure are most likely to directly affect patient care?

Hospitals may have as many as 50 committees, each responsible for various aspects of patient care and hospital management. Some of the more important ones are described below. If the patient has a concern involving an area over which one of these committees has responsibility, direct communication with the chairman or a member of the committee may bring appropriate action. A list of chairpersons and individual members can be obtained from the sponsoring department, for example, executive staff from the Dept. of Medicine; quality assurance from Administration; pharmacy and adverse drug reactions from Pharmacy. Some important committees are:

Research and Human Trials Committee (Institutional Review Board or IRB)—This committee reviews all proposed research involving human subjects in the hospital or related services. The review should occur before the research is instituted to assure that the scientific design is sound, the potential benefits outweigh the risks, and the informed consent of each patient involved will be obtained.

Utilization Review Committee—This committee is designed to review the appropriateness of the use of hospital services, including the length of hospital stay, and laboratory and diagnostic tests given, including blood work, X-rays, etc. Its purpose is to improve the use of hospital services, particularly by eliminating hospital stays that are too lengthy.

Quality Review Committee—This committee is established to evaluate the quality of patient care, to define and respond to problems in the delivery of good care. The committee reports the results of studies and related recommendations and has a largely advisory function.

Infection Control Committee—This committee studies data related to the incidence of infections among patients. Infection rates for the surgical staff in particular are tracked and investigated. The committee formulates policies and procedures related to infection control, including AIDS.

Nursing Policy and Procedures Committee—This committee establishes and reviews nursing department policies and procedures that govern the delivery of nursing care.

Institutional Ethics Committee (IEC)—This committee, a product of the 1980s, exists in most major hospitals. Its primary purpose is to raise the level of "ethical" discussion and awareness in hospitals. It may also set policy for particular areas (e.g., DNR orders and terminating treatment on incompetent patients). The committee is sometimes available for consultation, and if a patient is having a problem getting rights recognized because the attending physician (or anyone) feels it is "not right" or "unethical" to do what the patient requires, consulting the committee chairperson might be helpful in resolving this conflict.[6]

Where does the patient fit into the hospital hierarchy?

The dehumanizing aspects of institutional medical care have been well documented. An increased awareness by patients of the pressures to act in a certain way and an increased awareness of patient rights may help patients withstand such pressures and manage to maintain more dignity and autonomy. Talcott Parsons' famous formulation of the traditional patient role is:
1. The patient is not responsible for his condition and therapy is necessary for his recovery.
2. The patient is exempt from normal obligations.
3. Being sick is inherently undesirable, and the patient has an obligation to get well.
4. The patient and his family are obligated to seek competent help and to cooperate with the help.[7]

This formulation puts the patient in an unenviable position. If the patient is not responsible for various obligations, the patient is also not in the position of an autonomous moral agent. If competent help must be cooperated with, the patient is prone to paternalistic manipulation on the part of the medical staff. The patient's role is different for chronic conditions and conditions that are seen as self-imposed, and a particular individual may not conform to this model. Nonetheless, medical literature is filled with articles about patient compliance to therapeutic regimes, and medical professionals have strong feelings about what constitutes a desirable and undesirable patient. In teaching hospitals, it has been noted that "the stress of clinical training alienates the doctor from the patient [and] in a real sense the patient becomes the enemy. *(Goddamnit, did she blow her I.V. again? Jesus Christ, did he spike a temp?)*"[8] (emphasis in original). The writer who referred to patients as "the enemy" was a young anthropologist who had just completed his medical school training. At first, he thought this attitude merely an "inadvertent and unfortunate concomitant of medical training" but later came to see it as "intrinsic."

> Not only stress and sleeplessness but the sense of the patient as the cause of one's distress contributes to the doctor's detachment. This detachment is not just objective but downright negative. To cut and puncture a person, to take his or her life in your hands, to pound the chest until ribs break, to decide on drastic action without being able to ask permission, to render a judgment about whether care should continue or stop—these and a thousand other things may require something stronger than objectivity. They may actually require a measure of dislike.[9]

The notion that "dislike" of the patient is required to be an effective healer will strike most patients as bizarre and frightening. But even if this is the way things are, this does not mean that this is the way they must be.

Patients may be treated as enemies, as machines whose bodily plumbing becomes coextensive with medical machinery, or as "pathologies" (the stroke patient or the gallbladder, for example). Hospitals tend to support dehumanization by aggregating specialized and highly technical services and by bureaucratization (formal rules, routinized jobs, excessive paperwork). But consumer pressure, the move for national health insurance, and other changes are working against dehumanization. Jan Howard suggests a model of ideal humanized health care: Patients are seen as unique, irreplaceable whole persons, inherently worthy of concern, autonomous beings who share in decision making and have a reciprocal or egalitarian relationship with their health care providers, who respond to patients with empathy and warmth.[10] Vigorous enforcement of patient rights can help make this model a reality.

How is the doctor–patient relationship defined in the hospital?

There are a number of formulations of the doctor–patient relationship. The major ones are:

1. *Activity–passivity.* The physician is active and the patient passive in decision making. This is analogous to the relationship of parent to child.
2. *Guidance–cooperation.* This is a much more common model. The patient interacts with the doctor but idolizes the doctor and generally obeys orders.
3. *Mutual participation.* This model postulates a more equal doctor–patient relationship with the patient taking care of himself or herself to the extent possible and participating in decision making.[11] This "partnership" model is promoted by the acknowledgment of patient rights.

Not all patients may want a more egalitarian relationship, and the responsibility of shared decision making may be scary for some. But few, if any, patients actually want to be treated like children. The relationship with one's doctor can almost always be changed for the better by an open discussion of what the patient wants and of mutual duties and obligations. Philosopher John Ladd has argued that a meaningful relationship between physician and patient is an integral part of medical ethics. The appeal to legal rights or legalistic reasoning of the type "you can't do that because it is immoral or illegal," he has reasonably suggested, might be best reserved for when this relationship has broken down and legal appeals are the patient's primary source of protection.[12] It is, of course, much easier to take this approach with a physician the patient has known for some time than with a hospital-based specialist or resident that the patient has just met.[13] Nonetheless, the effort is worth making, and most physicians will react positively. If they do not, firm insistance on one's rights is an appropriate and reasonable response.

Might a patient be seen as a troublemaker for asserting his rights?

Of course. Making legal appeals runs a risk. Judith Lorber's study of "good" and "problem" patients makes this point.[14] Her work bears out other studies indicating that hospital rules are for the benefit of staff, not patients, and that doctors and nurses act so as to reduce patient autonomy and encourage compliance. The chief method of minimizing troublemaking by patients is to withhold information that patients might complain about.

Lorber says that most patients resent submission, but feel that submissiveness is the proper posture, lest their care suffer. Her study was done in a general hospital, using elective surgery patients. Although younger and better-educated patients tended to resist conforming attitudes and behavior more, most of the patients tended to comply with orders unquestioningly (only 24 percent of patients disagreed with their doctors at least once). Patients who were cooperative, stoic, and uncomplaining were labeled "good patients" by staff, while those demanding time and attention thought to be disproportionate to their medical condition were the "problem patients." The staff did expect patients to make them aware of their needs. Nonetheless, those labeled "problem patients" were more likely to be tranquilized or discharged early.[15]

Patients do sometimes pay a price for exercising their right to decide their own fates. Yet, autonomy requires accepting responsibility for one's action, even if adverse consequences result. Without increased pressure on hospitals to recognize patient rights, those who do exercise their rights will continue to be seen as troublemakers, and the system will not change.

NOTES

1. *See* American Hospital Association, *Hospital Statistics* (Chicago: AHA, 1987). For an excellent history of the American hospital, see C. Rosenberg,

The Care of Strangers (New York: Basic Books, 1987); and see T. Christoffel, *Health and the Law* (New York: Free Press, 1982) at 105–40.

2. J. K. Galbraith, *The New Industrial State* (Boston, Mass.: Houghton Mifflin, 1967), at 176. As Galbraith notes, the goal of expansion "in the output of many goods is not easily accorded a social purpose. More cigarettes cause more cancer. More alcohol causes more cirrhosis. More automobiles cause more accidents, maiming and death, more preemption of space for highways and parking, [and] more pollution of the air and the countryside" (*id.* at 164).

3. Knowles, "The Hospital," *Scientific American*, Sept. 1973, at 143.

4. New York regulations restrict house staff to twelve-hour emergency room shifts, and twenty-four-hour shifts in other areas of the hospital without time off, to a maximum of eighty a week on average (Barrow, "Making Sure Doctors Get Enough Sleep," New York Times, May 22, 1988, at E7). In addition to reduced working hours, hours must be better distributed, ancillary services improved, moonlighting restricted, and noncompliance with new rules punished. *See* McCall, *No Turning Back: A Blueprint for Residency Reform*, 261 JAMA 909–10 (1989); and Colford & McPhee, The Ravelled Sleeve of Care, 261 JAMA 889 (1989).

5. "A Look Ahead: What the Future Holds for Nursing," RN, Oct. 1987, at 101-8; "Fed up, Fearful and Frazzled," *Time*, Mar. 14, 1988, at 77-78.

6. *See generally* R. Cranford & A. E. Doudera, eds., Institutional Ethics Committees and Health Care Decision Making (Ann Arbor, Mich.: Health Administration Press, 1984).

7. T. Parsons, "Definitions of Health and Illness in the Light of American Values and Social Structure," in E. G. Jaco, ed., *Patients, Physicians and Illness* (New York: Free Press, 1958).

8. M. Konner, *Becoming a Doctor* (New York: Penguin, 1988), at 373. *See also* P. Klass, *A Not Entirely Benign Procedure* (New York: Putnam, 1987; S. Hoffman, *Under the Ether Dome* (New York: Scribner's, 1987; and S. Shem, *House of God* (New York: Marek, 1978).

9. Konner, *supra* note 8, at 373; and see *supra* note 4.

10. J. Howard, "Health Care," *Encyclopedia of Bioethics* (New York: Free Press, 1978) II: 619-23.

11. *See* S. Gorovitz, *Moral Problems in Medicine* (New York: Prentice-Hall, 1976).

12. J. Ladd, "Legalism in Medical Ethics," in J. Davis, ed., *Comtemporary Issues in Biomedical Ethics* (Clifton, N.J.: Humana Press, 1978). *See also* Callahan, *Contemporary Biomedical Ethics*, 302 New Eng. J. Med. 1228 (1980).

13. *See*, e.g., Okrent, "You and the Doctor: Striving for a Better Relationship," New York Times Magazine, Part II, Mar. 29, 1987, at 18.

14. Lorber, *Good Patients and Problem Patients: Conformity and Deviance in a General Hospital*, 16 J. Health & Social Behavior 213-25 (1975).

15. Id. *And see* R. Macklin, *Mortal Choices* (Boston, Mass.: Houghton Mifflin, 1987, at 214–16.

III

Rules Hospitals Must Follow

Hospitals are not isolated islands, and physicians are not foreign diplomats with legal immunity. Medicine must be practiced within the framework of the law of the United States and of the state in which the patient, physician, and hospital are located. All health care facilities are bound by state regulations, by the standards of the Joint Commission on Accreditation of Healthcare Organizations (JCAH) if they are so accredited, and by their own internal bylaws and policies. Public facilities, in addition, must assure that those human rights guaranteed by the United States Constitution are afforded to all patients. *Perhaps the most important thing for patients and their advocates to realize is that few doctors and nurses have a sophisticated understanding of what the law is,* and often when they think they know the law, they have it wrong. Readers of this book will almost certainly know more about the law than anyone a patient is likely to encounter in a health care facility, HMO, or doctor's office.

Until the mid-1960s, the medical profession had almost complete authority to set standards and regulate medical practice in the hospital. In the past two decades, however, courts have found the hospital legally responsible for the conduct of physicians on a variety of theories.[1] This has led to a movement by hospital administrators and their trade organization, the American Hospital Association (AHA), to achieve a measure of control over medical decision making. If the hospital can be held responsible for the actions of doctors, the administrators properly reasoned, they have the right to exercise some control over the doctors. A major step in this direction of interest to patients was the promulgation of the AHA's Patient Bill of Rights.

This chapter examines some of the external and internal rules that health care facilities must follow under the law. A review of these materials reveals, unfortunately, that the combination of public policing and self-enforced rules is often insufficient to protect the rights of patients.

What state laws must a hospital follow?

Most of the state laws that directly relate to hospitals are found in the public health laws and hospital licensing laws of the state and in the regulations promulgated under the authority of these laws by the state agency (usually called the Department of Health) that has the duty to enforce them. All states now license hospitals, but prior to World War II, fewer than a dozen states had any laws regulating hospitals. It was the passage by Congress of the Hill–Burton Hospital Construction and Survey Act of 1946 that led most of the states to adopt a licensing statute, because licensing was a prerequisite of obtaining federal funds to build or expand hospitals. Since this was a construction bill, it should not be surprising that most of the standards set forth in licensing statutes pertain to architectural design and adequacy of the physical plant rather than the quality of care delivered in the hospitals. Moreover, even on these limited criteria, state enforcement has been generally lax, and "regulations are often worded as recommendations rather than requirements; words like sufficient, adequate, and reasonable are common, especially in standards dealing with patient care."[2] Even though states do have the power to directly affect the quality of care in hospitals, they have been loath to exercise it and have generally limited their activities to building and construction.[3]

What is the Joint Commission on Accreditation of Healthcare Organizations (JCAH)?

JCAH is the only private accrediting agency of national significance. It was founded as the Joint Commission on Accreditation of Hospitals in 1952 by the AMA, the AHA, the American College of Physicians, the American College of Surgeons, and the Canadian Medical Association (which has since formed its own group in Canada). Representatives of the American Association of Homes for the Aging and the American Nursing Home Association were added later, and the name was changed to its current name in 1988. The organization inspects health care facilities on a voluntary basis and designates them "accredited" if they measure up to a set of published capability criteria. Although submission to such an examination is optional, since 1965 a JCAH-accredited facility has been automatically eligible to be certified as a reimbursable provider under Medicare, provided it also complies with federal utilization review requirements. Until 1970, JCAH standards mainly dealt with the medical staff organization, record keeping, and the hospital's physical plant. New standards have since been added in areas such as the emergency department, anesthesia, nursing, infection control, and ambulatory care.

Unfortunately, the fact that a hospital has been accreditated by JCAH is no assurance that proper medical care is delivered in the hospital. Calling the JCAH "one of the most powerful and secretive groups in all of health care," a three-month 1988 *Wall Street Journal* investigation of JCAH concluded that "accreditation masks serious failings in possibly hundreds of the 5100 hospitals in America inspected and approved by the Joint Commission." The report found an inherent conflict of interest between the funding and control of JCAH (by the AMA and the AHA) and their inspection role and noted that the commission's concern is not patient-centered at all but hospital-centered: "The Joint Commission sees its business as helping hospitals...to stay in business, not regulating them in a legal sense or punishing them." This is why it almost never denies accreditation (failing only one in one thousand hospitals in the past two years), and rarely conducts surprise inspections or reviews patient-complaint files. The *Wall Street Journal* report concluded that "although the Joint Commission is the first to learn of a hospital's sloppiness, it is usually the last to act decisively."[4] Its president concedes that JCAH is not a regulatory group, but one that works on a cooperative, voluntary, and confidential basis with hospitals. He stated the organization's conflict of interest in these terms: "The Joint Commission has an obligation to protect the public, but also to be sure the survey results are valid and don't inappropriately embarrass the hospital."[5]

Some courts have reached a similar conclusion about JCAH. In one case, for example, five of eight dialysis patients at a JCAH-accredited hospital were negligently dialyzed with water that contained excessive amounts of aluminum. They suffered speech disturbances, seizures, walking disorders, and the potentially fatal disease of dialysis dementia. The patients settled a lawsuit with the state of Louisiana, which then sued the JCAH for negligence in its inspection. A Louisiana court of appeals found in favor of the JCAH, ruling that the JCAH surveys and accreditation are not done for the benefit or protection of hospital patients:

The duty to survey did not include within its scope the protection of hospital patients against injury from inattention and possible malpractice of certain hospital personnel and the malfunctioning of certain equipment...the purpose of the JCAH survey was to foster *self-improvement by the hospital.* Any benefit in favor of the hospital patients that resulted from the contractual agreement was merely incidental. JCAH did not contract to monitor, regulate, or supervise the standard of [the hospital's] patient care. In fact, JCAH had no authority to mandate compliance with accepted medical standards or to require remedial action.[6] (emphasis in original)

Since JCAH is not a patient-protection or regulatory agency, and since its record of protecting patients has been so poor, JCAH should be replaced with a public federal regulatory agency that will not only accredit hospitals but will also make its findings available to the public, make patient safety and protection its first priority, and go out of its way to solicit and investigate complaints against hospitals by patients and members of the public. Until such a public agency is created, patients and their advocates should insist that hospitals at least adhere to the JCAH standards on patient rights.

Which standards of the JCAH are most relevant to patient rights?

Patient rights are dealt with in a section of the *Accreditation Manual* that precedes the standards and is entitled "Rights and Responsibilities of Patients." It provides in relevant part:

Access to Care
Individuals shall be accorded impartial access to treatment or accommodations that are available or medically indicated, regardless of race, creed, sex, national origin, or sources of payment for care.

Respect and Dignity
The patient has the right to considerate, respectful care at all times and under all circumstances, with recognition of his personal dignity.

Privacy and Confidentiality

The patient has the right, within the law, to personal and informational privacy, as manifested by the following rights:

- To refuse to talk with or see anyone not officially connected with the hospital, including visitors, or persons officially connected with the hospital but not directly involved in his care.
- To wear appropriate personal clothing and religious or other symbolic items, as long as they do not interfere with diagnostic procedures or treatment.
- To be interviewed and examined in surroundings designed to assure reasonable visual and auditory privacy. This includes the right to have a person of one's own sex present during certain parts of a physical examination, treatment, or procedure performed by a health professional of the opposite sex and the right not to remain disrobed longer than required to accomplish the medical purpose for which the patient was asked to disrobe.
- To request a transfer to another room if another patient or a visitor in the room is unreasonably disturbing him by smoking or by other actions.

Identity

The patient has the right to know the identity and professional status of individuals providing service to him and to know which physician or other practitioner is primarily responsible for his care. Participation by patients in clinical training programs or in the gathering of data for research purposes should be voluntary.

Information

The patient has the right to obtain, from the practitioner responsible for coordinating his care, complete and current information concerning his diagnosis (to the degree known), treatment, and any known prognosis. This information should be communicated in terms the patient can reasonably be expected to understand.

Communication

The patient has the right of access to people outside the hospital by means of visitors, and by verbal and written communication.

When a patient does not speak or understand the predominant language of the community, access to an interpreter should be provided, especially when language barriers are a continuing problem.

Consent

The patient has the right to reasonably informed participation in decisions involving his health care. To the degree possible, this should be based on a clear, concise explanation of his condition and of all proposed technical procedures, including the possibilities of any risk of mortality or serious side effects, problems related to recuperation, and probability of success. The patient should not be subjected to any procedure without his voluntary, competent, and understanding consent or that of his legally authorized representative. Where medically significant alternatives for care or treatment exist, the patient shall be so informed.

The patient has the right to know who is responsible for authorizing and performing the procedures or treatment.

The patient shall be informed if the hospital proposes to engage in or perform human experimentation or other research/education projects affecting his care or treatment, and the patient has the right to refuse to participate in any such activity.

Consultation

The patient, at his own request and expense, has the right to consult with a specialist.

Refusal of Treatment

The patient may refuse treatment to the extent permitted by law.

Transfer and Continuity of Care

A patient may not be transferred to another facility without receiving a complete explanation of the need for the transfer and of the alternatives to the transfer, and unless the transfer is acceptable to the other facility. The patient has the right to be informed by the practitioner responsible for his care, or his delegate, of any continuing health care requirements following discharge from the hospital.

Hospital Charges

Regardless of the source of payment of his care, the patient has the right to request and receive an itemized and detailed explanation of his total bill for services rendered in the hospital. The patient has the right to timely notice prior to termination of his eligibility for reimbursement by any third-party payer for the cost of his care.

Hospital Rules and Regulations

The patient should be informed of the hospital rules and regulations applicable to his conduct as a patient. Patients are entitled to information about the hospital's mechanism for the initiation, review, and resolution of patient complaints.

Are these standards an integral part of the JCAH standards?

Yes, although JCAH has toned down some of its enthusiasm for patient rights. In the 1970s, the patient rights section concluded: "The spirit and intent expressed in this preamble relative to the hospital patient's rights and needs and the observance of these in practice will be considered as a persuasive factor in the determination of a hospital's accreditation, *in the manner as are any of the standards of this volume*" (emphasis added).

The 1989 manual states simply that the enumerated patient rights "are considered reasonably applicable to all hospitals."

What is the American Hospital Association's (AHA) Bill of Rights?

The AHA Bill of Rights was approved as a national policy statement after a three-year study by the AHA's board of trustees and four consumer representatives. When it was released to the press in early 1973, the AHA said that none of its 7000-member hospitals would lose accreditation if it failed to adopt the statement and make copies available to all patients, but it expected its members to endorse the statement.[7] Response over the past fifteen years has been mixed, but most hospitals have adopted a statement similar to that promulgated by the AHA.

What does the AHA Bill of Rights provide?

The text of the AHA Bill of Rights is:

1. The patient has the right to considerate and respectful care.
2. The patient has the right to obtain from his physician complete current information concerning his diagnosis, treat-

ment, and prognosis in terms the patient can be reasonably expected to understand. When it is not medically advisable to give such information to the patient, the information should be made available to an appropriate person in his behalf. He has the right to know, by name, the physician responsible for his care.

3. The patient has the right to receive from his physician information necessary to give informed consent prior to the start of any procedure and/or treatment. Except in emergencies, such information for informed consent should include but not necessarily be limited to the specific procedure and/or treatment, the medically significant risks involved, and the probable duration of incapacitation. Where medically significant alternatives for care or treatment exist, or when the patient requires information concerning medical alternatives, the patient has the right to such information. The patient also has the right to know the name of the person responsible for the procedures and/or treatment.

4. The patient has the right to refuse treatment to the extent permitted by law and to be informed of the medical consequences of his action.

5. The patient has the right to every consideration of his privacy concerning his own medical care program. Case discussion, consultation, examination, and treatment are confidential and should be conducted discreetly. Those not directly involved in his care must have the permission of the patient to be present.

6. The patient has the right to expect that all communications and records pertaining to his care should be treated as confidential.

7. The patient has the right to expect that within its capacity a hospital must make reasonable response to the request of a patient for services. The hospital must provide evaluation, service, and/or referral as indicated by the urgency of the case. When medically permissible, a patient may be transferred to another facility only after he has received com-

plete information and explanation concerning the needs for and alternatives to such a transfer. The institution to which the patient is to be transferred must first have accepted the patient for transfer.

8. The patient has the right to obtain information as to any relationship of his hospital to other health care and educational institutions insofar as his care is concerned. The patient has the right to obtain information as to the existence of any professional relationships among individuals, by name, who are treating him.

9. The patient has the right to be advised if the hospital proposes to engage in or perform human experimentation affecting his care or treatment. The patient has the right to refuse to participate in such research projects.

10. The patient has the right to expect reasonable continuity of care. He has the right to know in advance what appointment times and physicians are available and where. The patient has the right to expect that the hospital will provide a mechanism whereby he is informed by his physician of the patient's continuing health care requirements following discharge.

11. The patient has the right to examine and receive an explanation of his bill, regardless of source of payment.

12. The patient has the right to know what hospital rules and regulations apply to his conduct as a patient.

In 1985 the AHA's Special Committee on Biomedical Ethics concluded that no changes were needed in this document.

What is the legal significance of the JCAH standards, the AHA Bill of Rights, and other similar documents on patient rights?

The thrust of the AHA Bill of Rights is to encourage etiquette and courtesy. "Rights" like "considerate and respectful care," to "every consideration of privacy," to "reasonable response to requests of a patient for services," and to "expect reasonable continuity of care" are essentially matters of staff courtesy.

The remainder of the listed rights are all fairly simple concepts relating to firmly established legal rights derived from informed consent. The AHA list is, therefore, essentially a combination of rudimentary statements of courtesy and basic concepts concerning informed consent. This led Dr. Willard Gaylin to term the title "not only pretentious, but deceptive" and to describe the entire AHA effort as akin to "the thief lecturing his victim on self-protection."[8]

Herbert Dennenberg, at the time Pennsylvania's insurance commissioner, issued a Citizen Bill of Hospital Rights in response to the AHA document in April 1973 because he believed that the AHA document "omits some of the most basic rights." He asserted that "what patients need and must have a right to is competent and quality care at prices they can afford." The Pennsylvania statement urges changes in the entire health care system and goes beyond the narrow confines of the hospital by arguing for public access to hospital data and continuity of care. The final provision is labeled "consumer advocacy" and begins: "The public has a right to expect a hospital to behave as a consumer advocate rather than as a business headquarters for doctors and hospital officials. The hospital should affirmatively and aggressively move to protect the patient and his interests rather than rubber stamp the demands of doctors."[9]

Hospitals have a long way to go to equal strides taken in the areas of employer–employee, landlord–tenant, and debtor–creditor relationships. Nonetheless, the AHA document served important symbolic purposes and helped legitimize the entire patient rights movement.

In a survey of nurses, for example, adoption of a Patient Bill of Rights was listed as the most common way of changing hospital policy regarding patient rights. More important, nurses at hospitals where the bill of rights was taken seriously reported greater job satisfaction and pride in the quality of patient care.[10] If adopted as official hospital policy, a Patient Bill of Rights can also be used in court as evidence of a standard of care to which the hospital and its staff can be held. In some early cases, a few courts concluded that to permit the introduction of self-imposed standards into evidence would only discourage the use of such standards,[11] but the modern trend is to allow them into evidence. The JCAH standards and hospital bylaws, for example, have been admitted to serve as "evidence of custom" to aid the jury in determining the standard of care to which the hospital should be held.[12] A court might even allow the admission of a document like the AHA Bill of Rights, even if the defendant hospital had not officially adopted it, on the grounds that it would give the jury an indication of industry custom as practiced by other hospitals.

This same conclusion applies to the JCAH standards. Some JCAH statements are unequivocal and thus offer much stronger potential protection to patients than the vague AHA statements. Examples include the nondiscrimination policy statement, the right to wear appropriate personal clothing, the right to visitors, and the right to consult with a specialist.

Some states, including Arizona, California, Illinois, Kentucky, Maryland, Massachusetts, Michigan, Minnesota, New Hampshire, New York, Pennsylvania, Rhode Island, and Vermont have adopted a Patient Bill of Rights by regulation, resolution, or statute. In these states, hospitals are legally required to follow the standards set forth in these acts. These enactments are primarily a restatement and codification in one place of existing statutory and common-law rights of patients. Thus, citizens in other states possess most, if not all, of these rights as well. Nonetheless, formal statutory enactment of a Patient Bill of Rights is a powerful educational tool and can promote enforcement of patient rights in the state. A model Patient Bill of Rights Act is set forth in appendix B.

What is the general rule concerning the ability of a patient to sue a hospital for violation of a hospital rule or regulation?

In general, violation of a rule or regulation that is designed primarily for the safety of hospital patients is malpractice if the violation results in injury to the patient.[13] This rule was applied in one case in which a consultation was sought by the patient but refused.[14] The court cited the JCAH standards (since the hospital was accredited and had agreed to be bound by these standards) on consultation, which provided in part: "The patient's physician is responsible for requesting consultation when indicated. It is the duty of the hospital staff through its chief of service and executive committee to make certain that members of the staff do not fail in the matter of calling consultants as needed." A malpractice award was upheld as a result. In reaction, JCAH revised this standard by lowering it to state only that "the use of consultations, and the qualifications of the consultant, should be reviewed as part of medical care evaluation," and later dropped it altogether.

Can a health care facility be held liable for permitting a doctor it knows or should know is practicing substandard medicine to continue to practice medicine in the facility?

Yes. Perhaps the best known example is described by a lower-court California judge. In that case a doctor, John G. Nork, was found to have performed at least thirteen unnecessary laminectomies (operations on the spinal cord), some that disabled the patient for life. One patient who sued Nork and the hospital in which he operated had suffered multiple injuries because of unnecessary surgery performed on him. The court awarded $2 million in punitive damages as well as $1.7 million in compensatory damages to the patient. The hospital settled with the patient and the amount assessed against Nork was affirmed on appeal. Although this lower-court case is not precedent in any jurisdiction, and was reversed and ultimately vacated on appeal, this portion of the opinion is generally thought to have been correctly decided. In discussing the liability of the hospital, the court noted:

> The hospital has a duty to protect its patients from malpractice by members of its medical staff when it knows or should have known that malpractice was likely to be committed upon them. Mercy Hospital had no actual knowledge of Dr. Nork's propensity to commit malpractice, but it was negligent in not knowing...because it did not have a system for acquiring the knowledge; it did not use the knowledge available to it properly; it failed to investigate the Freer case [a previous malpractice charge against Nork].... Every hospital governing board is responsible for the conduct of their medical staff.[15]

In an analogous Arizona case, the court found that when a hospital had undertaken to monitor and review the performance of staff doctors and to restrict or suspend their privileges if they had demonstrated an inability to perform a certain procedure, it could be held liable for failure to take action against a physician who had demonstrated such inability.[16]

For what other acts might a health care facility be held liable?

Facilities can be held liable for the negligent acts of their employees (the doctrine of *respondeat superior*) and for failure to fulfill the duty the facility owes to its patients directly (the doctrine of *corporate responsibility*). The following examples illustrate areas in which a hospital may be liable for its action or inaction. The list is not meant to be exhaustive.

A hospital may be held liable for failure to require that instruments be counted after surgery, even if not counting them is common practice.[17] Likewise, failure of a hospital radiology department to obtain and check the medical history of a patient referred for X-ray may be negligent, even though it was common practice not to check.[18] A hospital violates good standards if it allows, because of lack of sterile technique, a patient or employee with a known infectious condition to infect a previously noninfected patient.[19]

Good practice requires a nurse or other hospital employee to immediately report defective or inadequate hospital equipment.[20] A nurse is also required to read the "signals of danger" and bring these to the immediate attention of a doctor,[21] and if the doctor persists in a course of action that may be to the detriment of the patient, the nurse has an obligation to report this in such a way as to ensure the patient's safety.[22] A facility may not transfer a patient if it will worsen the patient's condition,[23] must not injure a bedridden patient while turning the patient over,[24] nor permit a patient to be burned by a heating lamp, hot-water bottle, or similar device.[25]

How does the hospital staff view the patient?

The view the hospital staff will take of the patient varies from hospital to hospital. United States hospitals in the nineteenth century were heavily class-structured. Only the poor went to hospitals, and "every aspect of the patient's life was subject to the institution's paternal oversight."[26] Patients were expected to work in the wards to the extent their health permitted, no reading was allowed without permission, and use of profane language could result in discharge. Visitors were controlled by a "rigid system of passes."[27] Patients were, in short, treated like children.

By the 1930s and 1940s, many hospitals adopted the posture that hospital patients should be treated like guests, and the hospital should act like a hotel that rents space to patients and their physicians. This view continued into the 1950s and 1960s, but because it shut the hospital administration out of the medical care arena, it ultimately became untenable. Since then, as the law has made the hospital's duties to patients more explicit, the hotel model has withered. In the 1980s, as health care began to be seen by many as a commercial product, some hospitals came to view patients as consumers.[28]

With the advent of informed consent, and the view that it is good for both patients and providers for patients to be active participants in their care, a new partnership model is developing in which patients are seen as "partners" in care.[29] It has been suggested, correctly I believe, that this is what patients want.[30]

The partnership model is supported by patient rights, and the recognition that each patient is an individual with unique values and the right to make his or her own decisions. The partnership model also requires the hospital staff physicians to engage the patient as a unique individual. Although this may seem more difficult, the partnership model produces added professional satisfaction for physicians and nurses as well.[31] Patients should insist on being treated as partners, lest their caregivers mistakenly conclude that patients want to be treated like children, guests, or passive consumers.

NOTES

1. *E.g., Darling v. Charlestown Community Memorial Hospital*, 33 Ill. ad 326, 211 N.E.2d 253 (1965), *cert. denied*, 383 U.S. 946 (1966); *Steeves v. U.S.*, 294 F. Supp. 446 (1968); *Elam v. College Park Hospital*, 183 Cal. Rptr. 156, 132 Cal. App. 3d 332 (1982).
2. Worthington & Silver, *Regulation of Quality of Care in Hospitals: The Need for Change*, 35 Law & Contemp. Probs. 305, 309 (1970).
3. *See generally* Havighurst, *Regulation of Health Facilities and Services by Certificate of Need*, 59 Va. L. Rev. 1143, 1144 (1973); and C. Havighurst, ed., *Regulating Health Facilities Construction* (Washington, DC.: American Enterprise Institute, 1974).

4. Bogdanich, "Prized by Hospitals, Accreditation Hides Perils Patients Face," Wall Street Journal, Oct. 12, 1988, at 10.
5. Pinkney, "Newspaper Story Slamming Joint Commission Inaccurate and Unfair, Supporters Say," American Medical News, Oct. 28, 1988, at 12.
6. *State v. Joint Commission on Accreditation of Hospitals*, 470 So. 2d 169, 177-78 (La. 1985); *and see* Jost, *The Joint Commission on Accreditation of Hospitals: Private Regulation of Health Care and the Public Interest*, 24 B.C.L. Rev. 835 (1983).
7. New York Times, Jan. 9, 1973, at 1.
8. Gaylin, "The Patient's Bill of Rights," *Saturday Review of Science*, Mar. 1973, at 22.
9. "Citizen Bill of Hospital Rights," Pennsylvania Insurance Dept., Harrisburg, Pa. (Apr. 1973); partially reprinted in Quinn & Somers, *The Patient's Bill of Rights*, 22 Nursing Outlook 240, 241(1974).
10. Sandriff, "Bill of Rights Makes Honesty Easier," RN, Aug. 1978, at 43.
11. *E.g., Fonda v. St. Paul City Ry.*, 71 Minn. 438, 74 N.W. 166 (1898).
12. *E.g., supra* note l; and *Stone v. Proctor*, 259 N.C. 633, 131 S.E.2d 297 (1963); *Pederson v. Dumouchel*, 72 Wash. 2d 73, 431 P.2d 973 (1967); *Magana v. Elie*, 108 Ill. App. 3d 1028, 439 N. E. 2d 1319 (1982); *Hahn v. Suburban Hospital Assoc.*, 461 A.2d 7 (Md. Ct. App. 1983).
13. *Kapuschinsky v. United States*, 248 F. Supp. 732 (D.S.C. 1966); *Jackson v. Power*, 743 P.2d 1376 (Alaska 1987).
14. *Steeves v. United States*, 294 F. Supp. 446 (1968).
15. *Gonzales v. Nork*, No. 228566, slip op. at 194 (Super. Ct. Cal. Nov. 19, 1973), *rev'd and remanded on other grounds*, 60 Cal. App. 3d. 728, 131 Cal. Rptr. 717 (1976), *vacated*, 143 Cal. Rptr. 240, 573 P.2d 458 (1978). Portions of this opinion are reprinted in S. Law & S. Polan, *Pain and Profit* (New York: Harper & Row, 1978), at 215.
16. *Purcell v. Zimbelman*, 18 Ariz. App. 75, 500 P.2d 335, 341, (1972); *see also Joiner v. Mitchell Cty. Hospital Authority*, 125 Ga. App. 1, 186 S.E.2d 307 (1971), aff'd, 189 S.E.2d 412 (1972).
17. *Leonard v. Watsonville Community Hospital*, 47 Cal. 2d 509, 305 P.2d 36 (1957). On hospital liability, *see generally* J. Smith, *Hospital Liability* (New York: Law Journal Seminars Press, 1986); and Mulholland, *The Corporate Responsibility of the Community Hospital*, 17 Toledo L. Rev. 343 (1986).
18. *Favalora v. Aetna Casualty & Surety Co.*, 144 So. 2d 544 (La. Ct. App. 1962).
19. *Helman v. Sacred Heart Hospital*, 62 Wash. 2d 136, 381 P.2d 605. (1963). See also Rosencranz & Larvey, *Treating Patients with Communicable Diseases*, 32 St. Louis U.L.J. 75 (1987).
20. *Rose v. Hakin*, 335 F. Supp. 1221 (D.D.C., 1971).
21. *Valentine v. Societe Francaise De Bienfaisance Mutuelle de Los Angeles*, 76 Cal. App. 2d 1, 172 P.2d 359 (1946); *Duling v. Bluefield Sanitarium,*

Inc., 149 W. Va. 567,142 S.E.2d 754 (1965). A hospital may also be liable for negligence in connection with the preparation, storage, or dispensing of medications. *See Ball Memorial Hospital v. Freeman*, 245 Ind. 71, 196 N.E.2d 274 (1964).

22. *Goff v. Doctors General Hospital of San Jose*, 333 P.2d 29 (Cal. App. 1958); and Darling, supra note 1.

23. *Alden v. Providence Hospital*, 127 U.S. App. D.C. 214, 382 F.2d 163 (1963).

24. *St. John's Hospital & School of Nursing, Inc. v. Chapman*, 434 P. 2d 160 (Okla. 1967).

25. Annotation, Hospital's Liability for Injury to Patient from Heat Lamp or Pad or Hot Water Bottle, 72 A.L.R. 2d 408 (1960).

26. C. E. Rosenberg, *The Care of Strangers* (New York: Basic Books, 1987), at 117.

27. Id.

28. *E.g.*, Levin, "Hospitals Pitch Harder for Patients," New York Times, May 10, 1987, at B1.

29. President's Commission for the Study of Ethical Problems in Medicine and Biomedical and Behavioral Research, *Making Health Care Decisions* (Washington, D.C.: Government Printing Office, 1982).

30. MacStravic, *The Patient as Partner: A Competitive Strategy in Health Care Marketing*, 33 Health & Hospital Services Admin. 15 (1988).

31. *E.g.*, Douma, *Informed Patients and Physician Satisfaction*, 243 JAMA 2168 (1980); *and see* J. Katz, *The Silent World of Doctor and Patient* (New York: Free Press, 1984).

IV

Emergency Medicine

Emergency medicine, usually centered in a hospital emergency department, presents both health care professionals and patients with unique problems. In a typical physician–patient interaction, the patient voluntarily enters the relationship, the physician gets to know the patient and has an opportunity to discuss treatment options in detail with time for reflection. In an emergency department, the patient is brought in by ambulance, the police, or others, does not choose the physician, and there is rarely opportunity for detailed discussion of treatment options. The pain or fear that drives people to seek medical advice is often more extreme in the emergency setting. In some cases, the situation may be even worse: The patient may be under the influence of drugs or alcohol, or be suffering a severe psychotic episode and actually be hostile to the nurses and physicians who are trying to help.[1]

The homeless may be brought to the emergency room by police because they are intoxicated and there is nowhere else to take them. For this group, which society seems to have abandoned, emergency care offers no long-term solution and frustrates care for others who need it. As one astute physician–administrator has described it, "The emergency ward is what the church had been in the Middle Ages: A sanctuary for those with any form of disease: social, psychic, or somatic."[2]

The good news is that emergency medicine is now a recognized and valued specialty, and the individuals working in emergency departments are highly skilled and knowledgable. In fact, even if a person has a family physician, when the person suffers an emergency, the family physician will usually advise the individual or family to go to the nearest emergency room for treatment. And most physicians will directly refer patients who are suffering from acute emergency conditions in the doctor's office to the nearest appropriate emergency facility.

Must a hospital render service to someone who comes to an emergency department for treatment?

The general rule is that if the person is experiencing a medical emergency, a hospital with emergency facilities is legally required to provide appropriate treatment if it can, or a referral for appropriate treatment if it cannot provide the treatment itself. No completely satisfactory definition of an emergency has been formulated. In general, an emergency is an injury or acute medical condition likely to cause death, disability, or serious illness if not attended to immediately.

The physician's role is central. First, it is the physician's duty to determine whether or not an emergency situation exists. If an emergency does exist, both law and medical ethics require the physician to treat the patient or to find someone who can. The broadest definition of a medical emergency is that of the American College of Emergency Physicians, whose view is that a patient has made an appropriate visit to an emergency department when "an unforeseen condition of a pathophysiological or psychological nature develops which a prudent lay person possessing an average knowledge of health and medicine, would judge to require urgent and unscheduled medical attention most likely available, after consideration of possible alternatives, in a hospital emergency department." This definition may be motivated in part to satisfy third-party payers, but it is also one that most lay people would find reasonable. The group gives examples of such conditions, including relief of acute or severe pain; hemorrhage or threat of hemorrhage; and obstetrical crises and labor. All these conditions require a person to be seen by a physician. Once seen, "a true emergency is any condition clinically determined to require immediate medical care."[3]

What are some examples of emergency conditions?

Examples of emergency conditions that require the immediate attention of a physician include:

Massive hemorrhage from major vessels (heavy bleeding)
Cardiac arrest (heart stoppage)
Cessation or acute embarrassment of respiration (breathing stopped)
Profound shock from any cause (collapse with increased heart rate and pale skin tone)

Ingestion or exposure to a rapidly acting poison
Anaphylactic reactions (allergic response)
Acute epidural hemorrhage (collection of blood within brain fol-
    lowing head injury)
Acute overwhelming bacteremia and toxemia (release of bacteria
    into blood stream, decrease in blood pressure, increase in tem-
    perature)
Severe head injuries
Penetrating wound of the pleura or pericardium (heart or lung wound)
Rupture of an abdominal viscus (any internal organ of abdomen)
Acute psychotic states (sudden and complete change in personality)[4]

An emergency condition need not be this serious, however, and
could include broken bones, fevers, and cuts that require stitches.
The leading case explicitly dealing with the right to receive emergency treatment, for example, involved a four-month-old baby with
diarrhea and fever.[5] The family physician had prescribed medication by phone on the second day of the illness and saw the child
during office hours on the third day.

The child did not sleep at all the third night, and on the morning of
the fourth day, the parents, knowing their doctor was not in his office
that day, took their child to an emergency room. The nurse on duty
refused to examine the child, telling the parents, among other things,
that hospital policy forbade treating anyone already under a doctor's
care without first contacting the doctor, which she was unable to do.
The parents took their child home and made an appointment to see
their family doctor that night. The child died of bronchial pneumonia
later that same afternoon.

The court ruled that the parents could recover from the hospital
for refusal to treat an "unmistakable emergency." The court's reasoning was that people should be able to rely on an "established custom" of
the hospital to render emergency assistance in its emergency room.
One who requires emergency assistance but is refused treatment at a
place where there was a reasonable expectation that treatment would
be rendered is worse off by having been delayed in obtaining
treatment. The court applied the principle that if one voluntarily under-
takes to render services, service cannot be refused or negligently

terminated to the detriment of one who has reasonably relied on the representation that service would be provided.[6]

A state can also impose a duty on hospitals to administer emergency care by statute or regulation.[7] In emergencies, no distinction is made between the duty of private and public hospitals, except that a public hospital may be held to a higher standard.[8] No hospital may refuse treatment on the basis of the prospective patient's race, religion, or national origin.[9] And no hospital may refuse to treat a patient who has an emergency condition because the patient is infected with HIV (human immunodeficiency virus) or has AIDS.[10]

What are the COBRA rules?

In April 1986, Congress adopted the Consolidated Omnibus Budget Reconciliation Act of 1985 (COBRA).[11] An amendment to COBRA established criteria for emergency services in hospitals that admit Medicare patients (almost all do) and criteria for safe transfer of patients between hospitals. Under this federal law:

1. A hospital must provide all patients with a medical screening examination to determine whether an emergency medical condition exists;
2. A hospital must provide stabilizing treatment to any individual with an emergency medical condition or woman in active labor prior to transfer;
3. If the hospital cannot stabilize the patient, the patient may be transferred to another hospital:
   a. if the responsible physician certifies in writing that the benefit of the transfer outweighs the risk;
   b. if the receiving hospital has space and personnel to treat the patient and has agreed to accept the patient;
   c. if the transferring hospital sends medical records with the patient; and
   d. if the transfer is made in appropriate transportation equipment with necessary life support.
4. If a hospital knowingly and willfully, or negligently violates any of these provisions, it can be terminated or suspended from the Medicare program; and

5. If a physician or hospital knowingly violates the law, a civil monetary penalty of $25,000 can be imposed on each of them for each violation of the law.

The Health Care Financing Administration (HCFA) has the primary responsibility for enforcing the law, and the Office of Inspector General of the US Department of Health and Human Services (HHS) is responsible for applying most of the penalties.

By early 1988, it became clear that the Reagan administration was not interested in enforcing this law, that almost no one was being told about it, and that no meaningful enforcement procedures were being followed.[12] Indeed, the administration even refused to adopt regulations to enforce the statute until Congress issued a report entitled "Equal Access to Health Care: Patient Dumping" in March 1988.[13] That congressional report documented the continued problem of patient "dumping" in the United States, using independent studies of the problem and specific examples of failure of the federal government to take action under the COBRA provisions. For example, a St. Louis hospital was found not in compliance with the Act, but no action was taken against the hospital. A pregnant teenager had come to the hospital in May 1987 complaining of abdominal pain and vaginal bleeding. Prior to being transferred to another hospital, she went into labor and delivered a stillborn five-month-old fetus. The hospital nonetheless proceeded with the transfer even though the umbilical cord was not cut and the placenta not delivered. Primarily because few persons knew where to complain, fewer than two hundred complaints were received in the first two years of the Act.[14]

Not until June 1988 did HHS finally issue proposed regulations to enforce it.[15] It should be stressed that COBRA provides that a private malpractice action can be initiated by anyone who suffers harm resulting from the hospital's violation of the law. Because private attorneys seldom accept malpractice cases in the absence of severe injury, it has been suggested that the statute be amended to include the award of attorney fees for successful lawsuits. This would be very helpful to all persons seeking emergency services.[16] Until effective enforcement is forthcoming, it is critical for those who accompany people in emergencies to emergency rooms to insist (loudly if necessary) that they be treated (and not transferred except for better care), and that the law be complied with.

What kinds of emergency departments are there?

The Joint Commission for Accreditation of Healthcare Organizations issues standards for four different levels of emergency services, although in reality many hospitals have hybrid systems. Standards can apply to all services or be limited to one or more, for example, surgery and/or obstetrics. It is important to know the location of the Level I and Level II emergency wards nearest you, but probably more important to know how the ambulance emergency system (often a "911" phone number) functions in your area. Ambulance attendants (emergency medical technicians, EMTs) are usually well trained to make decisions regarding the appropriate emergency facility. It is also very helpful if you have a personal physician who is usually reachable by phone (and/or who is a member of a group that provides coverage for each other) so that your medical history can be quickly available to the emergency room physicians.

All JCAH-accredited hospitals must have at least one of the following levels of emergency care:

1. *Level I.* A Level I emergency department/service offers comprehensive emergency care 24 hours a day with at least one physician experienced in emergency care on duty in the emergency care area. There is in-hospital physician coverage by members of the medical staff or senior-level residents for at least medical, surgical, orthopedic, obstetrical/ gynecological, pediatric, and anesthesiology services. Other specialty consultation is available within approximately 30 minutes; initial consultation through two-way voice communication is acceptable.
2. *Level II.* A Level II emergency department/service offers emergency care 24 hours a day with at least one physician experienced in emergency care on duty in the emergency care area, and with specialty consultation available within approximately 30 minutes by members of the medical staff or senior-level residents. Initial consultation through two-way voice communication is acceptable.

3. *Level III.* A Level III emergency department/service offers emergency care 24 hours a day with at least one physician available to the emergency area within 30 minutes through a medical staff call roster. Initial consultation through two-way voice communication is acceptable. Specialty consultation is available by request of the attending medical staff member or by transfer to a designated hospital, where definitive care can be provided.
4. *Level IV.* A Level IV emergency service offers reasonable care in determining whether an emergency exists, renders lifesaving first aid, and makes appropriate referral to the nearest facilities capable of providing needed care. A mechanism for providing physician coverage at all times is defined by the medical staff.[17]

How long must a patient wait to be seen?

The patient should be seen within a reasonable time, and in case of an emergency this means very quickly. In one case, the patient entered the emergency room bleeding from a shotgun wound in his arm. He was observed but not treated for two hours, at which time he was transferred to another hospital. He died shortly after arrival. The court understandably found the hospital responsible for the results of the delay in rendering care.[18]

The American College of Surgeons has recommended, "Medical staff coverage should be adequate to insure that an applicant for treatment will be seen by a physician within fifteen minutes after arrival."[19] This standard, however, seems much more ideal than real, and emergency room physicians believe that it can raise unrealistic expectations. A reasonable triage system, which includes an evaluation by a nurse or other health care professional, will ensure that those persons requiring immediate treatment (for example, chest pain, bleeding, major trauma) to safeguard their life or health will obtain it. The majority of emergency patients, however, do not require immediate treatment, and it is not uncommon to have to wait from one to six hours to have their condition treated. It is important to realize that patients are not seen on a first-come, first-served basis, but that the sickest and most critical patients are seen first.

Does a patient have a right to be seen by a physician?

There is a legal right to be screened by competent triage personnel, and if the patient is determined to have a medical emergency, the patient has a right to be seen and examined by a physician. Both the federal government in the *Medicare Conditions of Participation* and the American College of Surgeons in their *Standards for Emergency Departments in Hospitals* go further and specify that "every applicant for treatment should be seen by a physician." Initial screening or triage will usually be done by a nurse but is occasionally done by a clerk. This person determines the need for immediate medical intervention and decides how long the patient can reasonably wait to see a physician.

A hospital that operates an emergency department must, to be accredited, have an "emergency medical evaluation or initial treatment [that] is properly assessed by qualified individuals." [20] Although this standard does not specify the profession of the "qualified individuals," what case law there is indicates that if there is reasonable basis for suspecting an emergency exists, a patient has a right to be examined and treated by a physician. In one case, for example, an automobile accident victim was brought in with back pain. A nurse examined him and found no injury. She refused to call a doctor or admit the patient. The following day, in another hospital, he was found to have a broken back. The first hospital was held liable. [21]

Similarly, in another case, the patient was taken to an emergency ward with chest pain and shortness of breath. He told the nurse on duty he thought he was having a heart attack. Because he did not have the proper type of health insurance, however, she refused to admit him or call his physician to the hospital. The patient returned home and died, and the court decided that the hospital could be found liable for the death. [22]

After an emergency patient begins receiving treatment, is a hospital required to continue treatment?

Generally, a hospital must continue to provide treatment until a patient can be transferred or discharged without harm. In one case, a woman with a stab wound was examined and had her wound cleansed and dressed by an intern. She was then transferred to another hospital where she died a short while later during exploratory surgery. The court found that the hospital had not supplied adequate emergency treatment prior to ordering a transfer, and that this contributed to the patient's death.[23]

Another case involved a victim of an auto accident. After pulse and blood pressure checks and a brief abdominal exam, the intern in charge left the person unattended. After about forty-five minutes, the patient was transferred to another hospital, where he died thirty minutes later from internal injuries. The court found that the hospital had failed to provide adequate emergency treatment.[24]

Perhaps the best known "transfer" case involved Michael Thompson, a thirteen-year-old who was pinned against a wall by an automobile that had fallen off a jack and was rushed by ambulance from the scene directly to the Boswell Memorial Hospital in Sun City, Arizona.[25] He arrived at 8:22 PM and was examined and treated by the emergency room physician, who found that Michael's left thigh was severely lacerated; there was no pulse in the leg; the left foot and toes were dusky, cool, and clammy; and the bone was visible at the lower end of the laceration near the knee. The physician administered fluids, ordered blood, and called in an orthopedic surgeon.

The surgeon examined Michael, consulted by phone with a vascular surgeon, determined that Michael's condition was "stabilized," and thus "medically transferable." At 10:13 PM, Michael was placed in an ambulance and transferred to the County Hospital where his condition worsened. His condition later stabilized, and surgery was performed at about 1:00 AM. He survived but had serious residual impairment of the left leg, caused by the delay in restoring blood flow, which stopped when the femoral artery was cut.

The hospital stipulated that the surgery could have been performed at Boswell, and that the transfer was "for financial reasons." A Boswell administrator testified that emergency "charity" patients are transferred from Boswell to County whenever a physician, in his professional judgment, determines that "a transfer could occur." The emergency room physician did so determine, and a witness for the plaintiff testified that the physician told Michael's mother, "I have

the shitty detail of telling you that Mike will be transferred to County."
His mother "begged" the doctor not to send her son there. The primary
question before the court was whether the hospital violated the law
in transferring this child solely because he lacked the proper insurance.
The court reaffirmed Arizona law that "as a matter of public policy,
licensed hospitals in this state are required to accept and render care
to all patients who present themselves in need of [emergency] care.
The patient may not be transferred until all medically indicated emergency care has been completed...without consideration of economic
circumstance."

The court concluded that the child had an emergency condition,
that emergency surgery was the indicated treatment, and accordingly,
that the hospital had a legal duty to provide that emergency surgery.[26]

What is the physician's role in transfer?

The physician's role is pivotal, since no patient will be transferred
unless a physician certifies that the patient is stable and transferable.
If immediate treatment is necessary, it must be provided. If the patient
can be safely transferred without any foreseeable risk to health or
any decrease in the chance for recovery, the patient can legally be
transferred. But what if the county hospital (or tertiary care center)
refuses to take the patient? The physician can either arrange for the
patient to stay at the original hospital or be discharged. In this regard,
an Arizona court held that "since cessation of hospital care may not
be medically indicated despite the cessation of the emergency
condition...the private hospital may not simply release a seriously
ill, indigent patient to perish on the streets."[27] The hospital's obligation
to provide care after the emergency condition is stabilized continues
until the patient is properly transferred or is medically fit for discharge
from the hospital, and the physician should ensure that the hospital
meets this patient care obligation.

There may be some cases, especially in smaller, community hospitals, where the hospital does not have the properly trained staff, either
on site or on call, to stabilize the patient prior to transfer to a facility
that does have physicians (e.g., thoracic surgeons or training in the
use of intraaortic balloons) to treat the patient properly. In this case,

a decision to transfer must be made based on the physician's medical judgment of what course of action is in the patient's best interests.

May a hospital condition emergency care on prepayment or demonstration of an ability to pay?

No. Both COBRA and case law require emergency treatment to be rendered regardless of ability to pay. In one case, for example, an eleven-year-old boy was taken to a hospital for an emergency appendectomy.[28] Two hours later, after having been placed in bed and given medication, he was discharged by the hospital because his mother could not immediately pay $200. Although the court properly treated the case as an issue of negligent discharge rather than as one of refusal to administer emergency aid, the case is an example of a court finding that essential hospital treatment must be administered without regard to cost or ability to pay.

What does the term "dumping" mean?

Dumping refers to the practice of transferring undesirable patients to another facility. In his irreverent *House of God*, Samuel Shem describes gomers ("get out of my emergency room") as "human beings who have lost what goes into being human beings." They are elderly, demented patients, usually transferred from nursing homes, with multiple illnesses that medicine cannot cure. Law number one in the *House of God* was that "Gomers don't die." The problem of medical residents in the 1970s, according to Shem, was to keep such patients out of the emergency room (and therefore out of the hospital) because they would make your life miserable, and there was nothing you could do for them. In the 1980s, the incentives shifted. The new cry is to keep the uninsured and the poor, especially those with AIDS (the "new gomers"), out of the emergency rooms, based on economic law number one of the House of Adam Smith: "Poor People Can't Pay." The administrator of Dallas' Parkland Hospital has said that "dumping patients has been such a standard practice for so many hospitals that they don't even know they are doing something wrong."[29]

This disturbing development would have been completely unacceptable as recently as a decade ago. Although we remain the only industrialized nation other than South Africa not to have a system of national health insurance, we have always considered it the responsibility of our hospitals, as the purveyors of a social good, to provide emergency care to those in need regardless of ability to pay. As one court put it in 1973, "It would shock the public conscience if a person in need of medical emergency aid would be turned down at the door of a hospital having emergency service because that person could not at that moment assure payment for the service." [30] Efforts to transform medical care from a social good to an economic good threaten to erode this community ethic.

Princeton University economist Uwe Reinhardt has argued that because hospitals can no longer shift costs from one segment of their patient population to another, "the uninsured poor themselves [have] become the hot potatoes one hospital seeks to dump into the lap of another." [31] This transformation has not occurred without warning. For example, in 1980 a St. Louis man, stabbed in the back with a steak knife that was wedged against his spine, was transferred from an emergency room because he was uninsured, and the hospital refused to take the knife out unless he could come up with $1000 cash in advance. [32] Changes in Medicaid rules produced a significant increase in patient "dumps" to Cook County Hospital in Chicago in late 1983. [33] Also in 1983, New York passed a statute aimed at curtailing economic emergency room refusals in response to the deaths of two heart attack victims, who died after emergency treatment was refused. The statute makes denial of emergency care a misdemeanor punishable by a $1000 fine and a year in jail for the doctor, nurse, or other hospital personnel involved. [34]

Dumps are usually made within the same city or state, but not always. The farthest dump in the United States on record involved a Florida hospital that chartered a plane to fly a terminally ill AIDS patient to San Francisco to die. A spokeswoman for the hospital said the proposal for such transport was made by the physician in a conference with social workers who were seeking outpatient discharge. [35] The patient died sixteen days after transfer.

It is not just the rise of for-profit medicine that has challenged our traditional social commitment to provide emergency services to rich and poor alike but the erosion of this social commitment on the part of government itself.[36] And it is not just public hospitals that are the recipients of economic dumps. As Emily Friedman, an early chronicler of the phenomenon, notes, "Catholic institutions in many cities, and children's hospitals in a few, appear to be receiving significant numbers of economic transfers."[37]

What can be done to help guarantee that people experiencing a medical emergency will be treated promptly in a hospital emergency department?

Four actions seem reasonable:

• Professional associations should reaffirm the ethical requirement of their members to assist all those needing emergency medical care and should permit transfer only for better care.

• States, through statutes and regulations, should define "emergency" broadly (rather than narrowly) and should add criminal penalties for hospitals, physicians, and nurses in emergency departments that refuse such services.

• Uninsured individuals should be encouraged to carry cards that set forth the federal law and their state's law regarding emergency treatment; these cards should contain a form for the emergency room physician to sign certifying that no emergency condition exists if he or she refuses to treat them for what they consider an emergency situation, and that transfer can be accomplished without risk and will provide superior services should transfer be ordered.[38]

• COBRA should be amended to provide:

Hospitals should be required to provide written notification to all patients of their rights under COBRA.

No patient should be transferred until given an oral and written explanation why the transfer will result in better care for the patient.

Tertiary hospitals must be required to accept patients from small, unequipped hospitals when tertiary care is needed by the patient (to prevent "reverse dumping").

There should be a periodic, random review of hospital transfers by HCFA to ensure that the law is being followed.

Private attorney fees should be able to be awarded for cases in which plaintiffs prove a violation of the COBRA provisions.[39]

Other approaches should also be explored. One is to develop "emergency transfer protocols" and a mechanism to enforce them. State departments of public health could usefully adopt such protocols as regulations, with input from the hospitals in the state, and enforce them by retrospective review of individual emergency transfers. Noncomplying hospitals could be eliminated from emergency systems, such as a 911 network, fined, or their emergency department's license or permit could be revoked. Noncompliant physicians and nurses could also be disciplined through the offices of state physician and nurse licensing agencies.[40]

Policies could be adopted that require the informed consent of the patient prior to transfer, because the patient is the one who will be most affected by the decision.[41] Some prepaid health plans, such as health maintenance organizations (HMOs), also have overly restrictive policies regarding the use of hospital emergency rooms by their subscribers. If you are a member of a prepaid plan, you should know what its policy is regarding emergency treatment, and if it is too restrictive, make your concerns known to the plan and other subscribers and try to get the policy changed.

The only nonstatutory right to medical care that United States citizens have is the right to be treated in an emergency room for an emergency condition. During the 1970s, it seemed that this right was secure and could be the basis for expanding "the right to medical care" in the country. In the late 1980s, even this limited right is in danger of shrinking significantly. All those interested in fairness, equity, and a medical care system that at least responds to emergencies without first inquiring into the patient's finances must condemn this trend. Of course, hospitals should be paid for emergency services; but the fact that we have not yet worked out a payment mechanism that is universally acceptable is insufficient justification for physicians and hospitals to radically alter their traditional caring behavior by converting necessary emergency services into an economic commodity, available only to those able to pay. The emergency rule should remain: Treat first and ask about ability to pay later.[42]

NOTES

1. Sanders, "Unique Aspects of Ethics in Emergency Medicine," in K. V. Iverson et al., eds., *Ethics in Emergency Medicine* (Baltimore, Md.: Williams & Wilkins, 1986), at 9–12.
2. Knowles, "The Hospital," *Scientifc American*, Sept. 1973, at 136; *and see* Brook & Stevenson, *Effectiveness of Patient Care in An Emergency Room*, 283 New Eng. J. Med. 904, 907 (1970); Berarducci, Delbanco & Rabkin, *The Teaching Hospital and Primary Care*, 292 New Eng. J. Med. 615 (1975); and French, "Emergency Rooms Overwhelmed as New York's Poor Get Sicker," New York Times, Dec. 19, 1988, at 1, B4.
3. *Thompson v. Sun City Community Hospital*, 141 Ariz. 597, 688 P.2d 605 (1984).
4. Flint, *Emergency Treatment and Management*, 3d ed. (1965), at 88. Some of the leading textbooks on emergency medicine are P. Rosen *et al.*, *Emergency Medicine: Concepts and Clinical Practice*, 2d ed., 2 vols. (St. Louis, Mo.: Mosby, 1988); Roberts & Hedges, *Clinical Procedures in Emergency Medicine* (Philadelphia, Pa.: W. B. Saunders, 1985); Fleisher & Ludwig, *Textbook of Pediatric Emergency Medicine*, 2d ed. (Baltimore, Md.: Williams & Wilkins, 1988); and Haddad & Winchester, *Clinical Management of Poisoning and Drug Overdose* (Philadelphia, Pa.: W. B. Saunders, 1983).
5. *Wilmington General Hospital v. Manlove*, 54 Del. 15, 174 A.2d 135 (1961).
6. Other courts have arrived at the same conclusion. *See, e.g., Stanturf v. Sipes*, 447 S . W. 2d 558 (Mo. 1969), *Williams v. Hospital Authority of Han County*, 119 Ga. App. 626, 168 S.E.2d 336 (1969).
7. *E.g.*, Cal. Health and Safety Code sec. 1797 *et seq.* (West 1988); Ill. Ann. Stat. ch. 111 1/2, para. 86 (Smith-Hurd 1977); N.Y. Public Health sec. 2805(b) (McKinney 1985); Tex. Rev. Civ. Stat. Ann. art. 4438(a) (Vernon 1976) and art. 4438(f) Vernon 1985).
8. *Williams v. Hospital Authority, supra* note 6; *see also Jackson v. Power*, 743 P.2d 1376 (Alaska, 1987).
9. *Cypress v. Newport News Gen. & Nonsectarian Hospital Ass'n.*, 375 F.2d 648 (4th Cir. 1967).
10. *See, e.g.*, Annas, *Not Saints but Healers: The Legal Duties of Health Care Professionals in the AIDS Epidemic*, 78 Am. J. Public Health 844 (1988); Kelen *et al.*, *Unrecognized Human Immunodeficiency Virus Infection in Emergency Department Patients*, 318 New Eng. J. Med. 1645 (1988).
11. P.L. 99-272, codified, 42 USC sec. 1395(dd).
12. *See, e.g.*, "16 Hospitals under Investigation on Charges of Patient Dumping," *American Medical News*, Jan. 15, 1988, at 2, 12.
13. Committee on Government Operations, U. S. House of Representatives, 100th Cong., 2d Sess., *Equal Access to Health Care; Patient Dumping*,

Union Calendar No. 326, Mar. 25, 1988; *and see* "Report Assails Patient Transfers," *New York Times,* Mar. 30, 1988, at A20.

14. Tolchin, "U. S . Seeks to Require Treatment of all Hospital Emergency Cases," *New York Times,* June 18, 1988, at 1, 8.
15. *Hospital Responsibility for Emergency Care,* Proposed Rules, 53 Federal Register 2513-22527 aune 167 1988).
16. *Equal Access, supra* note 13, at 19, *and see* Note, *Preventing Patient Dumping: Sharpening the COBRA's Fangs,* 61 N.Y.U. L. Rev. 1186 (1986); and Enfield & Sklar, *Patient Dumping in the Hospital Emergency Department,* 13 Am. J. Law & Medicine 561(1988).
17. Joint Commission on Accreditation of Healthcare Organizations, 1989 *Accreditation Manual for Hospitals* (1988).
18. *New Biloxi Hospital v. Frazier,* 245 Miss. 185, 146 So. 2d 882 (1962).
19. *Bulletin of the American College of Surgeons* (May–June 1963), at 112. *But see* French, "New Emergency-Room Rules Seen Going Unheeded in New York City," *New York Times,* Jan. 4, 1989, at 1, B4
20. Joint Commission, *supra* note 17, at 29.
21. *Citizen's Hospital Assoc. v. Schoulin* 48 Ala. App. 101, 262 So. 2d 303 (1972).
22. *O'Neill v. Montefiore Hospital,* 11 A.D.2d 132, 202 N.Y.S.2d 436 (1960).
23. *Jones v. City of New York,* 134 N.Y.S.2d 779 (Sup. Ct. 1954), *modified,* 286 A.D.2d 825, 143 N.Y.S.2d 628 (1955).
24. *Methodist Hospital v. Ball,* 50 Tenn. App. 460, 362 S.W.2d 475 (1961). *See also Mulligan v. Wetchler,* 39 A.D.2d 102, 332 N.Y.S.2d 68 (1972).
25. *Thompson v. Sun City Community Hospital,* 141 Ariz. 597, 688 P.2d 605 (1984).
26. *Id.*
27. *St. Joseph's Hospital v. Maricopa Co.,* 142 Ariz. 94, 98, 688 P.2d 986, 990 (1984).
28. *LeJeune Road Hospital v. Watson,* 171 So. 2d 202 (Fla. Dist. Ct. App. 1965); *Tabor v. Doctors Memorial Hospital,* 501 So. 2d 243 (La. Ct. App. 1987) .
29. Taylor, "Cracking Down on Patient Dumping," *National Law J.,* June 6, 1988, at 1, 30.
30. *Mercy Medical Center v. Winnebago Cty.,* 58 Wis. 2d 260, 268, 206 N.W.2d 198, 201 (1973).
31. Editorial, "Health and Hot Potatoes," Washington Post, Mar. 16, 1985. *And see* Ansberry, "Despite Federal Law, Hospitals Still Reject Sick Who Can't Pay," *Wall Street Journal,* Nov. 29, 1988, at 1, All (approximately 250,000 people are "dumped" from hospitals annually for economic reasons).
32. Signor, "Shifted from Hospital to Hospital with Knife in His Back," *St. Louis Post-Dispatch,* Oct. 26, 1980.
33. Schumer, "Hospitals Unload Poor on County," *Chicago Tribune,* Jan 3, 1984. *And see* Relman, *Economic Considerations in Emergency Care: What Are Hospitals For?* 312 New Eng. J. Med. 372-73(1985); Himmelstein et al., *Patient Transfers: Medical Practice as Social Triage,* 74 Am. J. Public Health 494–97 (1984).

34. Barbanel, "Refusing Patients Becomes a Crime," New York Times, Aug. 6, 1983. *And see* Barron, "Hospitals Get Orders to Reduce Crowding in Emergency Rooms," New York Times, Jan. 24, 1989, at 1, B4.
35. Tokarz, "Odyssey of AIDS Victim Ends in Death," *American Medical News*, Nov. 4, 1983. *See also People v. Flushing Hospital*, 122 Misc. 2d. 260, 471 N.Y.S.2d 745 (1983).
36. Wing, *Medicare and President Reagan's Second Term*, 75 Am. J. Public Health 78–84 (1985).
37. Friedman, "The Dumping Dilemma: The Poor Are Always with Some of Us," *Hospitals,* Sept. 1, 1982, at 53.
38. Such a card has been used for years by Legal Services of Middle Tennessee [800-342-3317], and even when it does not help procure needed services, it identifies the physician who determined that such services were not necessary.
39. *Equal Access, supra* note 13, at 20-21.
40. G. J. Annas, *Judging Medicine* (Clifton, N.J.: Humana Press, 1988), at 42-45.
41. Kellermann & Ackerman, *Interhospital Patient Transfer: The Case for Informed Consent*, 319 New Eng. J. Med. 643 (1988).
42. Annas, *Your Money or Your Life: Dumping Uninsured Patients from Hospital Emergency Wards*, 76 Am. J. Public Health 74 (1986).

V

Admission and Discharge

This chapter concentrates on the legal problems at the beginning and end of a patient's hospital stay: admission and discharge. Discharge is sometimes called "release" (for example, "he was treated and released"), but this term makes the hospital seem more like a prison than a medical care setting. Problems seldom occur when the patient's physician has made formal arrangements in advance for the admission or discharge of the patient. The physician is the critical player in a patient's admission and discharge from the hospital, generally making all the arrangements for both. But in the absence of a specific agreement between the patient's physician and the hospital (for example, a patient who seeks admission alone or who wishes to leave before the doctor wants the patient to leave), many legal issues can be relevant. This chapter deals with the most frequently occurring problem areas. For legal issues involved with the admission of an emergency patient, see the preceding chapter.

Should a person routinely seek a consultation or a second opinion before entering the hospital or agreeing to elective surgery?

Yes. Almost everyone now agrees that hospitals can be dangerous places, and that people should only go there when it is absolutely appropriate and medically necessary. For example, patients may be injured by physicians (iatrogenic injury), become infected in the hospital (nosocomial infections), and be given the wrong drugs. In one study of eight hundred consecutive admissions to a major hospital, for example, more than one-third developed a major or minor iatrogenic injury, which in fifteen of these patients directly contributed to the patient's death.[1] Almost two million patients annually acquire infections in hospitals, and thousands of these patients die from them.[2] And an average-sized, three hundred-bed hospital, will conservatively experience 11,000 medication errors annually, or more than thirty medication errors daily.[3] Errors are also commonplace in surgery and can lead to death.[4] Because surgery has been overutilized in the United States, and because it is dangerous, many insurance companies *require* their subscribers to get a second opinion for elective surgery. This is good advice. Information on surgery and second opinions is contained in Chapter VII, "Surgery."

What restrictions may a hospital place on the admission of a nonemergency patient?

The general rule is that the nonemergency patient has no right to be admitted to a hospital.[5] This is another way of saying that there is no universal legally enforceable right to hospital care in the United States.[6] Teaching hospitals may encourage the admission of cases that are "interesting" for their staffs and discourage the admission of more routine cases.[7]

This does not mean that a hospital may place any condition it pleases on patient admission. Private hospitals may screen admissions based on a number of criteria, but *they may not refuse admission on the basis of race, religion, or national origin.*[8]

May a hospital refuse admission on the basis of inability to pay?

This has been termed "one of the most hotly contested issues" in hospital admissions.[9] Although there is no general "right to hospital care" or admission, the use of financial ability as a limitation on admission has been severely reduced by statutes, regulations, and judicial decisions. As discussed in detail in the preceding chapter, for example, patients experiencing a medical emergency have a legal right to be treated (and admitted to the hospital if continuing treatment is required) regardless of ability to pay. In the case of a nonemergency, the rights of patients to admission depend largely on the specific hospital involved.

For example, if the hospital obtained financial assistance for construction under the Hill-Burton Act of 1946 (and most did), then the hospital is required to provide a reasonable volume of services to those unable to pay for it. Federal regulations (promulgated in 1979) require that such hospitals provide the lesser of 3 percent of the facility's operating costs for the past fiscal year or 10 percent of all federal assistance received by the facility.[10] The regulations prohibit discrimination based on any factor other than medical need. For example, a rule limiting admission to patients whose private physicians have staff privileges at the hospital is not permissible.

The Hill-Burton requirement to accept some indigent patients continues for twenty years after Hill-Burton-assisted construction is completed. Most important to prospective patients, notices about the availability of such funds must be published in the newspapers and posted in conspicuous places in the hospital. Notice must also be given to individual patients seeking services, and these notices must clearly set forth the eligibility requirements and the method to apply for care. A determination of eligibility must be made within two working days from the date of the request and provided to the patient in writing. Failure to advise the patient of the hospital's Hill-Burton obligation may preclude the hospital from suing the patient to collect the hospital bill.[11]

There are two categories of patients who might qualify for such assistance: those below the poverty level for the past year; and those above the poverty line by no more than double this amount. Provision of uncompensated services to this second category is at the option of the facility. Hill-Burton hospitals must also participate in Medicare (the federal program for those over sixty-five) and Medicaid (the state program for certain categories of individuals).

Many states have laws that prohibit denial of admission to hospitals based solely on inability to pay,[12] and a number of courts have also ruled (based on hospital charters and public policy) that public hospitals have a duty not to discriminate against the medically indigent in admissions.[13] It is also illegal for a hospital to require a deposit from a Medicare or Medicaid patient.

Some hospitals, nonetheless, do have policies requiring deposits from patients lacking private insurance, or who are not covered by Medicare or Medicaid. That such requirements still exist in these hospitals underlines the need for a system of universal entitlement to hospital care for all those who are in need of hospital services.

What steps must a hospital take before a patient is considered admitted?

The question is important because even though a hospital may not he duty-bound to care for a nonemergency patient, once the hospital has initiated treatment, it has a duty to continue to provide care until the patient can be safely discharged.

The old rule was that a duty to provide or continue needed treatment did not commence until the patient was *formally* admitted, which involved the actual completion of admission forms; an initial evaluation of the patient's condition was not sufficient action to constitute admission.[15] Courts, however have expanded the concept of admission to include many acts by hospital employees, so that virtually any act on behalf of the patient will constitute admission and assumption of the duty to treat.[16]

Two cases in which no forms were signed illustrate this point. In one, dressing the patient in a hospital gown for examination was found sufficient to constitute admission.[17] In another, a nurse's attempt to contact a doctor to care for the patient was sufficient to constitute assumption of hospital responsibility for the patient's health.[18]

What forms will a patient be asked to sign on admission to the hospital?

A patient will generally be asked to sign payment and insurance forms, a form concerning personal property brought to the hospital, a "blanket" consent form, perhaps a form designating someone else to make treatment decisions if the patient is unable to, and perhaps an organ donation form. The patient may, of course, refuse to sign any of these forms or may choose to modify them. If the admission is elective, however, and if insisting on the patient's signature on a particular form is reasonable and does not violate public policy, the hospital might refuse to admit the patient who refuses to sign.

Can a hospital lawfully prevent a patient from leaving?

No. A competent, adult patient of sound mind may leave the hospital at any time, and the hospital may not prevent it. If the hospital restricts a patient's freedom to leave, the hospital can be sued for *false imprisonment,* the intentional confinement of the patient by threat or physical barriers against the patient's will. No actual damages need be proved, since the law assumes harm to the patient from this conduct.

This rule applies even if the patient has not paid the bill. In one case, for example, a patient was detained for eleven hours for not paying her bill. In concluding that she could sue the hospital for false imprisonment, the court said: "The fact that the bill...had not been paid afforded no sort of excuse for detaining the [patient] against her will."[19] This rule also applies to detaining infants and children when their parents have not paid the bill. Courts have found hospitals liable for interfering with the parent's right to a child's custody when discharge has been refused because of failure to pay a bill.[20]

A hospital may prevent a person of unsound mind from departing if the person is a danger to his or her own life or the lives and property of others, provided the hospital follows the state's laws governing involuntary commitment of the mentally ill, usually including a determination by a qualified psychiatrist that the person is mentally ill and a danger to himself or herself or others.

Hospitals may also request patients to sign a "discharge against medical advice form," sometimes termed discharge "AMA." Patients, however, have no obligation to sign such a form as a precondition to release, and hospitals can protect themselves equally well by simply having the physician or nurse document the circumstances of the patient's departure in the medical record, as well as any noncoercive attempts to persuade the patient to stay. In rare cases, such as contagious diseases, the public health authorities may have power to quarantine a patient, thus restraining the patient from leaving the premises. The state's quarantine authority is spelled out in each state's public health laws and is generally very broad.

What is meant by the charge that patients are now being discharged "quicker and sicker?"

Since the mid-1980s hospitals participating in the Medicare program have been reimbursed for their services to Medicare patients on a predetermined fee scale based on the patient's admitting diagnosis. These categories are termed "diagnosis-related groups," or simply DRGs.[21] In order to give hospitals an incentive to cut costs, hospitals are paid a fixed rate for each patient. If the patient stays longer than the DRG rate in which the patient falls, then the hospital will "lose money" on the patient. On the other hand, if the patient is discharged prior to the time the money allotted by the DRG rate is used, the hospital will "make money."

DRGs, however, were never intended to set a standard of care or a guideline for the number of days any individual patient would stay in the hospital. Rather, they are based on the assumption that in a large group of patients with a particular diagnosis, the DRG payment rate will represent the "average" patient. There are also provisions for hospitals to obtain additional payment for "outliers," patients who need an unusual amount of hospitalization for their diagnostic classification. *Thus, there is no DRG rate for any particular patient, and it is incorrect to state that any patient has "used up their DRG."* Patients can be properly discharged only when the attending physician has determined, in the exercise of reasonable medical judgment, that the patient can safely be transferred to a different care facility (for example, a rehabilitation center or nursing home) or can safely return home. Although DRGs do not affect physician income, a 1988 survey of physicians by the American Society of Internal Medicine found that two-thirds of them felt under pressure to release patients earlier than they thought was medically appropriate.[22]

Must a patient leave the hospital if a utilization committee says that hospitalization is no longer "needed"?

A patient must leave the hospital only if the patient's attending physician agrees with the assessment of the *utilization review committee.* Under federal Medicare regulations, each hospital must establish a utilization committee to make a determination of whether each hospital admission is necessary and, if so, to review it after a specified period of time to determine if hospital care is still needed. If the committee determines that an extended stay is not appropriate, the attending physician is notified and given an opportunity to explain the reason for the continued hospitalization.

The attending physician should receive written notice of the determination within two working days of the admission (in the case of an unnecessary admission) or within two days of the end of an extended-stay period. If the committee's decision is not reversed, its effect is that no federal funds can be paid to the hospital for continuing the patient's hospitalization. The decision does not, however, mean that the patient must be discharged. As previously noted, discharge is dependent on the medical judgment of the physician and the will of the patient. Either the patient or health care provider can also seek review of the decision by filing a written request with the local Social Security office or directly to the United States Department of Health and Human Services.[23]

Does the patient's attending physician retain the responsibility regarding discharge even if there is no money to pay for continuing hospitalization?

Yes. The general rule is that no patient may be discharged except by written order of a doctor familiar with the patient's condition. If the patient disagrees with the order, the patient has a right to demand a consultation with another physician before the order is carried out. In any event, however, *the decision to discharge* must be made on the basis of the patient's medical condition and *may not take into consideration the patient's nonpayment of medical bills.*

In one case, for example, a private hospital admitted an eight-year-old boy suffering from osteomyelitis (a bone infection). The bill for treatment was $1000, but the boy's father could pay only $349. The hospital discharged the boy for failure to pay his bills, and instructed the father that the boy would be safe at home under the care of a physician. In fact, the physician was not able to provide proper care at home, and the boy's condition worsened. The court found the hospital liable for negligent and wrongful discharge.[24]

A California case further illustrates the physician's responsibility. A woman, who was a member of California's Medicaid program (MediCal), complained to her physician about circulatory problems in her legs.[25] He recommended surgery, and his recommendation was approved by a before-the-fact utilization review committee that approved ten days of postoperative hospitalization for Medicaid payment. The day before she was scheduled for discharge, the physician requested an eight-day extension because of severe complications that had developed. The reviewing physician, however, authorized only a four-day extension. The patient was discharged after this four-day period without any further requests for extensions or reconsideration. Nine days later, the patient was readmitted on an emergency basis, and her right leg had to be amputated. She sued the Medicaid program, arguing that it was negligent in not granting her physician's request for the eight-day extension, and that her premature discharge directly caused the amputation of her leg. The California Court of Appeals ruled that the discharge decision was not for Medicaid to make but for her attending physician. The question of payment is distinct (although not wholly unrelated) from the one of medical judgment for the health of the patient. The patient might be able to sue her physician but not Medicaid, which was only trying to control costs in a reasonable manner. In the court's words, although "cost consciousness has become a permanent feature of the health care system, it is essential that cost limitation programs not be permitted to corrupt medical judgment."[26]

The lesson from this case (and others) is clear: The attending physician's medical judgment is the most important factor regarding discharge. If patients do not think they are ready to go home or be transferred, their first recourse is to discuss the reasons for this feeling directly with the attending physician. If the discharge is medically indicated, and the patient still refuses to leave, the hospital may take steps to forcibly remove the patient as a trespasser. All steps must, however, be reasonable, and only the minimum amount of force necessary may be used.

Does a doctor ever have a duty to refer the patient to a specialist or to seek a consultation with a specialist?

When a patient is in a teaching hospital, he or she is generally seen by many physicians, including house staff and specialists. In such a situation, consultations and referrals take place almost as a matter of course. In nonteaching hospitals, however, the patient may have only one physician who is responsible for the patient's care. In this situation, questions concerning the duty of the attending physician to obtain a consultation or to refer the patient to a specialist may be of critical importance to the patient and the patient's family.

The general rule is that if a doctor knows, or should know, that the patient's ailment is beyond the doctor's knowledge, technical skill, ability, or capacity to treat with a likelihood of reasonable success, the doctor is obligated to either disclose this to the patient or advise the patient of the necessity of other or different treatment.[27]

Not only must the general practitioner consult a specialist, but the specialist must also consult a specialist in another field when indicated. If a reasonably careful and skillful attending physician would have suggested consulting a specialist, a doctor may be found negligent for not making such a suggestion.[28] In a circumcision case, for example, an infant's penis had to be amputated. Evidence showed that prior to the infant's discharge from the hospital, a black spot appeared on his penis, and that the spot continued to grow when he was home. On returning to the hospital the following day, both a pediatric specialist and a urology specialist were consulted. They concluded that the spot was caused by gangrene, but that it was too late to do anything about it. The court found the evidence sufficient to warrant a finding by the jury that the physician who performed the circumcision should have called in a specialist prior to the child's discharge from the hospital, when something might have been done to save the child's penis.[29]

Does the hospital have a duty to ensure that doctors practicing in the hospital seek and obtain indicated consultation?

Yes. Perhaps the best known case dealing with this issue was decided in Illinois in 1965.[30] In that case, an eighteen-year-old broke his leg while playing college football. He was treated by a general practitioner in the hospital's emergency ward. The doctor applied a plaster cast to the leg. Over the next two weeks, the young man almost continuously complained of severe pain and pressure. His leg turned a grayish color, and the smell in his room was terrible. He was finally taken out of the hospital by his parents, and his leg, having been found to be gangrenous, had to be amputated. The doctor settled out of court, and the case continued against the hospital. One of the major issues considered by the court was whether the hospital itself could be found negligent for failure "to require consultation with or examination by members of the hospital surgical staff skilled in such treatment; or to review the treatment rendered to the plaintiff and to require consultants to be called in as needed." The court concluded

from the evidence—which included the state's public health licensing regulations, the JCAH standards, and the bylaws of the hospital—that the jury could properly have found that the hospital was
negligent in failing to require a consultation, and a judgment against
the hospital was accordingly affirmed.

It is likely that when the need for consultation should be apparent
to the doctor, in view of the patient's worsening condition, both the
doctor and the hospital will be found liable for failure to ensure that
such consultation is obtained.[31]

Must a doctor refer a patient to a specialist or seek a consultation on request by a patient?

The AMA's Principles of Medical Ethics require physicians to
respond to such patient requests. Although this is an ethical, not a
legal guideline, refusal of such a patient's request "is an invitation
to a malpractice suit if the attending physician turns out to be wrong
[in suggesting to the patient that a consultation is not necessary]."[32]

One case in which a request for consultation was refused involved
an eleven-year-old boy who was brought to a United States Air Force
dispensary with severe abdominal pain and vomiting. The doctor
did some tests and referred the patient to the United States Naval Hospital with a diagnosis of "possible appendicitis." At the emergency department of the hospital, the boy was seen by another doctor who did only a
cursory physical examination. He concluded it was not appendicitis
and asked that the child be taken home. The child's mother asked that
another physician be called in to look at the child, but the doctor refused.
Early the following morning, the child started crying and rolling around
in extreme pain. He was again taken to the emergency department
and seen by an intern who gave the child pain medication and sent
him home. The next day, the boy returned to the dispensary. After
delays both there and at the hospital, surgery was finally performed
but only *after* the boy's appendix had ruptured. As a result of the
delay, peritonitis developed, and the child not only had to spend
an additional three weeks in the hospital, but also suffered permanent internal damage. Citing the AMA Principles, the
hospital's emergency policy, the United States Government's
Medical Department Manual, and JCAH standards, the court

concluded that both doctors in the emergency department, and therefore the hospital, could be found negligent for failure to seek consultation.[33]

A physician refusing a patient's request for a consultation or referral does so at his or her own peril. If the physician is wrong in reassuring the patient that the treatment is proper, the physician may be found negligent for failure to respond appropriately to the patient's request.

Who is responsible for paying for a consultation or a second opinion?

Like almost everything else that happens to a patient, the patient is responsible for paying the bills. Almost all insurance policies, however, cover medically indicated consultations, since they are recognized as an integral part of good medical care.

May a doctor refuse to continue to see a patient without obtaining the services of another doctor for the patient?

The general rule is that once a doctor–patient relationship is established, the doctor must continue to see the patient until one of the four following conditions is met:

1. It is terminated by the consent of both doctor and patient.
2. It is revoked by the patient.
3. The doctor's services are no longer needed.
4. The physician withdraws from the case *after* reasonable notice to the patient.[34]

Abandonment occurs when the physician unilaterally severs the doctor–patient relationship without the consent of the patient at a time when continued care is needed. If injury results to the patient because of the abandonment, the patient may successfully sue the doctor for damages. If the treatment is in a critical stage at which abandonment might be harmful to the patient, nonpayment of bills by the patient cannot be used as a justification for refusal by the physician to extend further aid.[35]

A Virginia case illustrates how abandonment can occur. The doctor properly set a fracture and applied a cast to a child's leg. Yet, when the patient later complained of pain and the doctor was called, he failed to respond. The doctor then left town. When the child's parents were unable to locate the doctor, they called in another physician who cut the cast to relieve the pressure and noted evidence of an infection. When the doctor returned to town, he again refused to see the patient in the hospital and instead discharged the patient. At home, the pain continued, and another physician was called, who observed necrotic (dead) spots on the child's leg, which had to be amputated. In upholding a jury verdict for the patient, the court restated the general rule: "After a physician has accepted employment in a case it is his duty to continue his services as long as they are necessary. He cannot voluntarily abandon his patient. Even if personal attention is no longer necessary in the treatment of an injured limb, the physician must, if the case calls for it, furnish the patient with instructions as to its care." [36]

If a physician discharges a patient from the hospital, and the patient has a relapse and calls the physician for help, the physician must take steps to help the patient or face a charge of abandonment unless another physician has taken over the patient's care.

Can a patient recover damages from a physician or a hospital for premature discharge?

As two noted medicolegal commentators have stated, "It is uniformly held by the courts that the premature discharge of a patient constitutes abandonment." [37] These commentators also note that "unfortunately, instances of premature discharge abound in the reported cases." [38]

Premature discharge can be illustrated by a particularly gruesome case from the 1920s. The physician involved had performed an unskillful operation for a strangulated hernia. After an improper incision, he closed the wound without attempting to relieve the obstructed bowel and informed the patient that she was going to die. He then ordered the patient sent home in a hearse and refused thereafter to see her at home in spite of her calls. The Rhode Island Supreme Court had no difficulty finding that the physician was liable for abandonment. [39]

What information should the patient have prior to discharge?

No patient should be discharged without a discharge plan and clear instructions regarding follow-up self-care, what problems to be on the lookout for, and whom to call if any of these problems occur. Many doctors and hospitals have standard forms they give patients that explain such basic information, and patients should take this information seriously. A social worker may also be assigned to the patient to help make arrangements for placement in a nursing home or an extended-care facility or for a visiting nurse or other health aide to see the patient at home during recuperation.

NOTES

1. Steel *et al., Iatrogenic Illness on a General Medical Service at a University Hospital,* 304 New Eng. J. Med. 638 (1981).
2. Dixon, "Nosocomial Bacteremia: Etiology, Diagnosis, and Prevention," *Hospital Physician* (July 1985), at 17; Horan *et al.,* "CDC Surveillance Summaries 1986: Nosocomial Infection Surveillance, 19~," *Morbidity and Mortality Weekly Report: CDC Surveillance Summaries 1986,* 17SS-29SS.
3. N. M. Davis & M. R. Cohen, *Medication Errors: Causes and Prevention* (Philadelphia, Pa.: Stickley, 1981). *See generally* C. B. Inlander, L. S. Levin & E. Weiner, *Medicine on Trial* (New York: Prentice-Hall, 1988), at 12~53.
4. Couch *et al., The High Cost of Low-Frequency Events,* 304 New Eng. J. Med. 634 (1981). On second opinions and unnecessary surgery, *see* McCarthy & Wildner, *Effects of Screening by Consultants on Recommended Elective Surgical Procedures,* 291 New Eng. J. Med. 1331(1974). Specific procedures most frequently found unnecessary included hysterectomy, dilataffon and curettage, breast operations, and gallbladder removal.
5. This rule may have variations based on the type of hospital (e.g., governmental or private), the condition of the patient, and whether or not there is an existing contractual obligation to admit patients such as hospital employees. *E.g., Norwood Hospital v. Howton,* 32 Ala. App. 375, 26 So. 2d 427 (1946).
6. *See generally* Cantor, *The Law and Poor People's Access to Health Care,* 35 Law & Contemp. Probs. 901(1970).
7. E. Mumford, *Interns: From Students to Physicians* (Cambridge, Mass.: Harvard U. Press, 1970), at 179.
8. Some commentators have argued that these restrictions apply only to hospitals receiving federal funding (*e.g.,* Medicare) or engaged in "state action." As a practical matter, however, these qualifications apply to almost every hospital in the United States. *See, e.g.,* Civil Ri~lts Act of 1~4, Titles II and VI, 42 U.S.C.A. sec. 2000 and the regulations promulgated (thereunder), 45 C.F.R. sec. 80, as well as the Civil Rights Restoration Act of 1988. *See also Simkins v. Moses H. Cone Memorial Hospital,* 323 F.2d 959 (4th Cir. 1963), *cert. denied,* 376 U.S. 938 (1963); *Doe v. General Hospital of District of Co-*

lumbia, 313 F. Supp. 1170 (D.D.C. 1970). A related question involves the conditions a state can place on eligibility for public assistance. The US Supreme Court has invalidated as unconstitutional an Arizona statute that required a year's residency in a county as a condition to receiving nonemergency medical care at the county's expense. The court followed *Shapiro v. Thompson,* 394 U.S. 618 (1969), in finding that the statute created an "invidious classification" that impinged on the right of interstate travel by denying newcomers "basic necessities of life" (*Memorial Hospital v. Maricopa Co.,* 415 U.S. 250 [1974]).

9. Health Law Center, *Hospital Law Manual* (Rockville, Md.: Aspen Publishers, 1988). *And see Hunt v. Palm Springs General Hospital,* 352 So. 2d 582 (Fla. Dist. Ct. App. 1977).
10. *Fed. Register* 29372 (May 18, 1979). Codified in 42 C.F.R. sec. 124.501 *et seq. See American Hospital Association v. Schweiker,* 721 F.2d 170 (7th Cir. 1983); and Wing, *The Community Service Obligation of Hill-Burton Health Facilities,* 23 B.C.L. Rev. 577 (1982). The Hill-Burton Act is codified as Title VI of the Public Health Service Act. 42 U.S.C. sec. 291. *And see Newsome v. Vanderbilt University,* 653 F.2d 1100 (6th Cir. 1981).
11. *Hospital Center at Orange v. Cook,* 177 N.J. Super. 289, 426 A.2d 526 (A.D. 1981); *Cooper Medical Center v. Boyd,* 179 N.J. Super. 53, 430 A.2d 261 (A.D. 1981).
12. *Hospital Law Manual, supra* note 9, "Admitting and Discharge;" e.g., N.Y. Pub. Health Law sec. 2805(b) (McKinney, 1976) and Tex. Rev. Civ. Stat. Ann. art. 4438(a) (Vernon, 1975).
13. *E.g., Williams v. Hospital Authority of Hall County,* 119 Ga. App. 626, 168 S.E.2d 336 (1969).
14. 20 C.F.R. sec. 405.10 (Medicare).
15. *Birmingham Baptist Hospital v. Crews,* 229 Ala. 398, 157 So. 224 (1934).
16. L. Goldsmith, ed., *Liabilitiy of Hospitals and Health Care Facilities* (New York: Practising Law Institute [No. H4-2894], 1973), at 102.
17. *LeJeune Road Hospital v. Watson,* 171 So. 2d 202 (Fla. Ct. App. 1965).
18. *O'Neill v. Montefiore Hospital,* 11 A.D.2d 132, 202 N.Y.S.2d 436 (1960)
19. *Gadsden General Hospital v. Hamilton,* 212 Ala. 531, 532, 103 So. 553, 554 (1925); in another case recovery was denied because the court was not convinced that the patient's apprehension that force would be used to detain her was "reasonable" (*Hoffman v. Clinic Hospital,* 213 N.C. 669, 197 S.E. 161[1938]).
20. *Bedard v. Notre Dame Hospital,* 57 R.I. 195, 151 A.2d 690 (1959).
21. *See* Vladeck, *Medicare Hospital Payments by Diagnosis-Related Groups,* 100 Ann. Internal Med. 576 (1984); Phillips & Wineberg, *Medicine Prospective Payment: A Quiet Revolution,* 87 W. Va. L. Rev. 13 (1984); and Iglehart, *Early Experience with Prospective Payment of Hospitals,* 314 New Eng. J. Med. 1460 (1986).

22. Page, "Internists: DRGs Adversely Affecting Patient Care," *American Medical News*, Mar. 18, 1988, at 32. *See also* Schramm & Gabel, *Prospective Payment*, 318 New Eng. J. Med. 1681 (1988); McCarthy, *DRGs—Five Years Later*, 318 New Eng. J. Med. 1683 (1988); and Tolchin, "Shift on Medicare Expected to Hurt Hospitals in Cities," *New York Times*, Oct. 19, 1988, at 1.
23. *See generally* Mellette, *The Changing Focus of Peer Review Under Medicare*, 20 U. Rich. L. Rev. 325 (1986).
24. *Meiselman v. Crown Heights Hospital*, 285 N.Y. 389, 34 N. E.2d 367(1941). *Cf. Hicks v. U.S.*, 368 F.2d 626 (4th Cir. 1966). (Dispensary physician determined that patient had harmless instead of lethal disease without properly testing for the lethal possibility. With prompt surgery the patient would have survived; instead she was sent home and died from a high intestinal obstruction. In finding the dispensary physician liable, the court said: "By releasing the patient, the dispensary physician made his diagnosis final, allowing no further opportunity for revision, and this prematurely determined final diagnosis was based on an investigation not even minimally adequate." The court went on to determine that the premature discharge of the patient was the proximate cause of death.)
25. *Wickline v. California*, 183 Cal. App. 3d 1175, 228 Cal. Rptr. 661 (1986).
26. *Id.*, 228 Cal. Rptr. at 672.
27. *Manion v. Tweedy*, 257 Minn. 59, 65, 100 N.W.2d 124, 128 (Minn. 1957). A doctor does not, however, become liable for the negligent acts of another physician merely by recommending the specialist, *Dill v. Scaka*, 175 F. Supp. 26 (E. D. Pa. 1959); and *Mincey v. Blando,* 655 S.W.2d 609 (Mo. App. 1983).
28. O'Hern, "Duty to Refer to Medical Specialist," in *AMA, Best of Law & Medicine*, 1.968–70 (1970), at 25–26; *Graham v. St. Luke's Hospital,* 46 Ill. App. 2d 147, 196 N.E.2d 355 (1964).
29. *Valentine v. Kaiser Foundation Hospitals*, 194 Cal. App. 2d 282, 15 Cal. Rptr. 26 (1961).
30. *Darling v. Charleston Community Memorial Hospital*, 33 Ill. 2d 326, 211 N.E.2d 253 (1965), *cert. denied,* 383 U.S. 946 (1966).
31. *See generally* Southwick, *The Hospital's New Responsibility*, 17 Cleve.-Mar. L. Rev. 146 (1968); Note, *Hospital Liability—A New Duty of Care,* 19 Me. L. Rev. 102 (1967); Mueller, *Expanding Duty of the Hospital to the Patient*, 47 Neb. L. Rev. 337 (1968); see also Chapter III, "Rules Hospitals Must Follow."
32. Holder, "Referral to a Specialist," in *AMA, Best of Law & Medicine*, 1968–70 (1970), at 27.
33. *Steeves v. U.S.*, 294 F. Supp. 446 (D.S.C. 1968).
34. Comment, *The Action of Abandonment in Medical Malpractice Litigation,* 36 Tul. L. Rev. 834, 835 (1962); *and see Dillon v. Silver,* 520 N.Y.S. 2d

751, 134 A.D. 2d 159 (1987).

35. *Id.* at 841; *E.g., Becker v. Janinski,* 15 N.Y.S. 675 (App. Div. 1891); *Ricks v. Budge,* 91 Utah 307, 64 P.2d 208 (1937). The physician may, however, properly terminate the doctor–patient relationship for refusal to pay if he gives the patient sufficient notice for the patient to obtain other medical attention (*Burnett v. Layman,* 133 Tenn. 323, 181 S.W. 157 [1915]).

36. Vann v. Harden, 187 Va. 555, 565–66, 47 S.E.2d 314, 319 (1948).

37. J. Waltz & F. Inbau, *Medical Jurisprudence* (New York: Macmillan, 1971), at 146.

38. *Id.; see also Mucci v. Houghton,* 89 Iowa 608, 57 N.W. 305 (1894); *Reed v. Laughlin,* 58 S.W.2d 440 (MO. 1933); *Meiselman v. Crown Heights Hospital,* 285 N.Y. 389, 34 N.E.2d 367 (1941). Medicare's prospective payment system seems to be increasing the number of patients who die in nursing homes (now more than 20 percent), as hospitals have a financial incentive to transfer even terminally ill patients to nursing homes. Whether such transfers are medically appropriate has yet to be studied. *See* Sager et al., *Changes in the Location of Death After Passage of Medicare's Prospective Payment System: A National Study,* 320 New Eng. J. Med. 433 (1989).

39. *Morrell v. Lalonde,* 120 A. 435 (R.I. 1923).

VI

Informed Consent

Information is power, and because information sharing inevitably results in decision sharing, the doctrine of informed consent has helped transform the doctor–patient relationship. This is why informed consent is the most important legal doctrine in both the doctor–patient relationship and treatment in health care facilities. Not only is it important because of its implications for power and accountability, but it is also important because many of the other rights patients have are either derived from or enhanced by the doctrine of informed consent. The basic concept is simple: A doctor cannot touch or treat a patient until the doctor has given the patient some basic information about what the doctor proposes to do, and the patient has agreed to the proposed treatment or procedure. The overwhelming majority of Americans agree with this proposition, and the foundation on which it stands: People have a right not to have their bodies invaded without their approval because of their interests in bodily integrity and self-determination. Put more simply, it is the patient's body. The patient is the one who must experience the invasion and live with its consequences. There is no obligation to accept any medical treatment, and it is remarkable that anyone ever considered it acceptable practice to treat a person without that person's informed consent. Physicians have no roving mandate to treat whomever they believe may need their services.

As one court summarized the law at the dawn of the twentieth century: "Under a free government at least, *the free citizen's first and greatest right* which underlies all others—*the right to the inviolability of his person*, in other words, his right to himself, is the subject of universal acquiescence, and this right necessarily forbids a physician...to violate without permission the *bodily integrity* of his patient by a major or capital operation" (emphasis added).[1]

In the most important study of informed consent to date, the President's Commission for the Study of Ethical Problems in Medicine concluded that informed consent has its foundations in law and is an ethical imperative as well. It also concluded that "ethically valid consent is a process of shared decision making based upon mutual respect and participation, not a ritual to be equated with reciting the contents of a form that details the risks of particular treatments." Its foundation is the fundamental recognition "that adults are entitled to accept or reject health care interventions on the basis of their own personal values and in furtherance of their own personal goals." [2]

In his best-selling book, *Love, Medicine, and Miracles*, surgeon Bernie Siegel also underlines the importance of informed consent and shared decision making to enhancing the doctor–patient relationship and avoiding malpractice lawsuits: "Shared responsibility increases cooperation and reduces the resentments that often lead to malpractice suits. Second-guessing and recrimination are unlikely when decisions are based on a mutual assessment of what is right for the patient now, not on predictions about the unknowable future...The physician must remember that it's the patient who must make the choice and then *live with it*" (emphasis in original). [3]

This chapter explores the doctrine of informed consent, including where it comes from, what it means in practice, how patients can be sure they have been properly informed, and the right to refuse or place conditions on consent. Unique problems involving dying patients are dealt with in Chapter XII ("Care of the Dying"), and issues involving consent and refusal of treatment for children are discussed in Chapter VII ("Surgery"). Although many states have specific statutes on informed consent, [4] it is fair to say that the doctrine of informed consent is substantially similar across the country. Where it differs, there is a trend toward requiring more, rather than less, disclosure and toward measuring the doctor's obligation by reference to what patients need to know to make their own decisions rather than by what doctors typically tell their patients. The goal of the doctrine is to enhance and encourage a responsible patient–physician partnership in treatment decisions.

What is the difference between consent and informed consent?

Traditionally, the unauthorized performance of a medical or surgical procedure was dealt with by the law of "battery," a legal term for any intentional, unauthorized, offensive touching. Examples could include a punch in the face, an unwanted kiss in the dark, or a push down the stairs. Some courts use the terms "assault," "battery," and "assault and battery" interchangeably. In one famous opinion, Judge Benjamin Cardozo wrote: "Every human being of adult years and sound mind has a right to determine what shall be done with his own body; and a surgeon who performs an operation without his patient's consent commits an assault for which he is liable in damages." [5]

Since battery connotes an unauthorized touching, it is most applicable either when the doctor treats a patient without obtaining any consent (for example, a young child incapable of understanding) or when the doctor properly obtains consent for one type of an operation but does another (for example, when a doctor operates on the wrong leg). The modern trend is to discard the battery model for all but these glaring types of misconduct (because in most cases treatment is in fact authorized) and to consider the doctor's lack of complete disclosure to the patient under the rubric of negligence.

Under the negligence theory, the doctor has an affirmative duty to the patient, based on the fiduciary, or trust, nature of the doctor–patient relationship, to disclose relevant, material facts about, and risks of, treatment. If these are not disclosed, and if the patient suffers one of these risks, the patient may sue the doctor for negligence. The distinction between suing in battery or negligence may seem only semantic. But in many states, a patient is still required to present expert evidence of a physician to prove how much the "average doctor of good standing" discloses (since in these states customary medical practice defines how much the doctor should have told his patient). In a battery lawsuit the failure to disclose risks and alternatives may render the consent meaningless, and any offensive touching therefore becomes a battery. In this view, expertise becomes irrelevant, since the doctor's privilege to touch the patient ends when he or she exceeds the scope of the patient's consent. [6] The choice between battery or negligence may also determine the period of time during which the lawsuit can be brought (statute of limitations), the nature and extent of the damages that can be recovered, and the test used to prove that the patient's injury was caused by the physician.

What is the doctrine of informed consent?

The doctrine of informed consent, simply stated, is that *before* a patient is asked to consent to any treatment or procedure that has risks, alternatives, or low success rates, the patient must be provided with certain information. This information includes at least the following, which must be presented in *language the patient can understand*:

1. A description of the *recommended treatment* or procedure
2. A description of the *risks and benefits* of the recommended procedure, with special emphasis on risks of death or serious bodily disability
3. A description of the *alternatives*, including other treatments or procedures, together with the risks and benefits of these alternatives
4. The likely results of *no treatment*
5. The *probability of success*, and what the physician means by success
6. The major problems anticipated in *recuperation*, and the time period during which the patient will not be able to resume his or her normal activities
7. Any other information generally provided to patients in this situation by other qualified physicians[7]

As can readily be seen, there is nothing profound or mysterious about it: It is based on common sense and what almost anyone would need to know to make a reasonable decision about whether or not to accept a treatment recommendation by a physician. Some physicians have argued that the doctrine puts them at a disadvantage, because it is difficult to determine what "risks" should be disclosed. The courts have generally answered as follows: The risks that must be disclosed are material risks; that is, those that might cause a reasonable patient to decide not to undergo the recommended treatment (and choose either an alternative treatment or no treatment at all). Another way to think about materiality or the importance of a risk is by multiplying the probability that a risk will materialize by the magnitude of that risk.[8] This implies that even a 1 in 10,000 risk of death must always be disclosed, but not a 1 in 10,000 risk of a two-hour headache.

All information must, of course, be presented in language the patient can understand, and treatment should not proceed until the health care provider is satisfied that the patient actually does understand the information presented.

Why have the courts adopted the doctrine of informed consent?

The doctrine of informed consent exists for two basic reasons: (1) to promote self-determination and (2) to promote rational decision making.[9] The reason that it is seen as reasonable to require physicians to provide this information is that patients are generally unlearned in medical sciences and so are "abjectly dependent" on their physicians for the information they require to make a personal decision about treatment.[10] Another way to think of it is that *caveat emptor* (let the buyer beware) is not an appropriate or useful model for the doctor–patient relationship. The patient *should* be able to trust that the doctor will share such critical information.

It has, however, been persuasively suggested that as courts view it, informed consent is too arbitrary a conceptionalization of the doctor–patient relationship, and that the reality is one of psycho-analytical richness, ambiguity, and uncertainty in which the patient often plays the part of the child, and the doctor that of the parent.[11] There is no doubt that there is much truth in this view, and that the doctor–patient relationship is much more complex than simple information exchange. Nonetheless, the parent–child relationship is one that few patients want with their doctor, and one the law should not encourage. All patients have a *right* to basic information even if it will not be useful to some patients, and sharing uncertainty can be as important as sharing other information. The law properly defines the disclosure rules doctors should be held to so that doctors know what is expected of them, and patients can obtain the basic information they need. This will not necessarily result in an ideal doctor–

patient relationship, but will help enhance patient autonomy and participation in decision making.

When must the patient's informed consent be obtained?

Almost always. The general rule is that any time there is an inherent risk of death or injury that the patient might not know about, when alternatives exist, or when the probability of success is low, the person performing the test or treatment is required to obtain the patient's informed consent. The rule applies equally to administration of drugs orally or by hypodermic needle,[12] the performance of diagnostic tests,[13] and the performance of major or minor surgical procedures.[14] An example of a procedure that probably does not require specific disclosures, because the risks are minimal and generally known, is the taking of a blood sample, although if an HIV test is being done, informed consent should be obtained.[15] Diagnostic tests like angiograms, bone marrow aspirations, and spinal taps also require informed consent.[16] The ethical requirement of informed consent is even broader, and no health professional should touch, treat, or discuss the patient without explaining why.

Is informed consent important in an outpatient clinic, HMO, or doctor's office?

As important as informed consent is to normal crisis-oriented medical treatment, it is even more critical to elective procedures, especially when there is a possibility that the condition might worsen as a result of the treatment. This point was illustrated in a case of a pregnant woman who had skin blotches on her face. She was treated by a physician who used a procedure called dermabrasion in which he, in effect, sanded a layer of skin off her face. Instead of removing the blotches, they actually became more noticeable. There was evidence that the doctor did not mention the possibility of failure and that he knew the probability of a good result was only about 50 percent. The court noted that this was "an elective thing," since "there was no emergency" and the "patient's health was not at stake." The doctor therefore was obligated to "disclose...all material facts which reasonably should be known if his patient is to make an informed and intelligent decision....Arguably, one of the facts needed...

was the percentage probability that the contemplated surgery would improve her appearance." The case was sent back for a jury determination of the sufficiency of the doctor's disclosures.[17]

Other examples of procedures that may produce results expressly contrary to those desired by the patient are plastic surgery[18] and vasectomies.[19] In all of these instances, patient knowledge of *the probability of success* is critical to providing informed consent to the treatment.

What does it mean to say that consent must be competent and voluntary?

In general, consent must be *competent, voluntary, informed, and understanding*. The last two elements have been discussed. "Voluntary" is relatively simple. The patient must not be under extreme duress, medicated, intoxicated, or threatened by the physician when giving consent. In one case, for example, a patient who had been given a sleeping pill was awakened in the middle of the night and asked to sign a consent form. The patient testified that he could not remember the event, and the court found that consent obtained under such circumstances was not valid.[20] The paradigmatic case for involuntary consent is when someone holds a gun to another's head and says, "Sign this."

Competence is somewhat more difficult to understand. The most important thing to know is that *the law presumes every adult is competent*. The burden of proof is on the person who would try to take an individual's right to make decisions away on the basis of "incompetence." The second most important thing to know is that a person is not incompetent simply because the person refuses treatment or disagrees with the physician. If this were so, the entire doctrine of informed consent would collapse into a "right to agree with the doctor." Nonetheless, almost the only time competence is likely to be challenged in the hospital is when a patient disagrees with a doctor's recommended treatment or refuses to be treated altogether.

The most useful way to think about competence in the health care setting is that an individual is competent to consent or refuse treatment so long as the individual understands the information needed to give informed consent for the proposed treatment. Relatives, friends, or even the physician can thus determine competence, because it is primarily a question of fact (even though only a judge can actually adjudicate an individual "incompetent" and appoint a guardian with legal authority to make decisions for the individual). Although there is no magic formula, one way to test competence is to determine if the patient can answer the following questions:

1. What is your present physical condition?
2. What is the treatment that is being recommended for you?
3. What do you and your doctor think might happen to you if you decide to accept the treatment? [This could be modified in appropriate circumstances to ask specifically about risks involved in the treatment, including those of most concern to the patient.]
4. What do you and your doctor think might happen to you if you decide not to accept the recommended treatment?
5. What are the alternative treatments available (including no treatment) and what are the probable consequences of accepting each?[21]

Since competence ultimately rests on an ability or capacity to understand and appreciate the nature and consequences of one's decision, it is appropriate to test this ability in the medical care setting by using a basic informed consent interview. In this, one would carefully explain the nature of the proposed treatment, its likely risks and benefits, the alternatives, their risks and benefits, the likely consequences of no treatment, and (ideally before the patient is asked to consent), would make a determination whether or not the patient actually understands this basic information. By making the competence determination *before* asking for consent, the "outcome approach" pitfall of labeling a patient incompetent solely on the basis of the patient's refusal is avoided.[22] The consent of the family, or anyone else, is, of course, neither necessary nor appropriate if the patient is competent.

Can a family member or the next of kin consent for the patient if the patient is incompetent?

There is a medical custom of deferring to the next of kin when treatment decisions concerning incompetent patients must be made. In cases in which there are no real treatment options, and treatment is in the best interests of the patient, this presents no major legal or ethical problems. Technically, however, only a guardian can make a legally binding treatment decision for an incompetent person. And a guardian can only be appointed by a judge, and only after the judge has determined that the individual is in fact incompetent.[23] Since guardians are required to make decisions consistent with the ward's "best interests," and since courts are likely to defer to the family's judgment of best interests in any event, there is often little to be gained by having a legal guardian appointed.[24] It is primarily when treatment that seems to be in the patient's best interests is being refused by the family or next of kin, or there is an unresolvable disagreement among family members, that it is appropriate to have a court-appointed guardian make the decision (*see* the discussion in Chapter XII).

One reason family members are often asked to consent on behalf of an incompetent person is simply that it is time consuming, expensive, and burdensome to have a guardian appointed. Moreover, by consenting to the treatment, the relatives effectively waive their own rights to sue the physician for any action based on failure to obtain consent. Their consent also demonstrates that the physician consulted someone likely to know and be concerned about the patient's well-being, and this will likely be sufficient to persuade the patient not to sue for failure to obtain consent should the patient recover.

When is a physician justified in not giving full information to the patient about a proposed treatment or procedure?

There are four major justifications:

1. If it is an emergency situation, where immediate treatment is needed to preserve the patient's life or health (*see* Chapter IV, "Emergency Medicine").
2. If the risks are minor and well known to the average person (such as the risks involved in drawing blood).
3. If the patient does not want to know the specific risks and, understanding that there are risks of death and of unspecified serious bodily disabilities, asks not to be informed of

them in detail (here the patient may be asked to sign a "waiver of informed consent" form, and this is perfectly appropriate).

4. The "therapeutic privilege" under which the physician need not inform the patient of the risks involved if the physician has objective evidence that this information would so upset the patient that the patient would be unable to make a rational decision. Under these circumstances, the physician must, however, give the information to another person designated by the patient. This privilege actually applies very rarely.

The California Supreme Court has put this fourth exception in the following terms: "A disclosure need not be made beyond that required within the medical community when a doctor can prove by a preponderance of the evidence that he relied on facts that would demonstrate to a reasonable man the disclosure would have *so seriously upset the patient that the patient would not have been able to dispassionately weigh the risks* of refusing to undergo the recommended treatment" (emphasis added).[25]

A better legal and ethical rule would be that the physician must always inform a competent patient of all material information, but that the manner in which the information is conveyed (including the time, place, and language used) could be permitted to vary, depending on the patient's circumstances.

Some doctors have complained that they might be held liable for telling the patient too much, but no court has ever held a doctor liable for giving his patient too much accurate information. On the other hand, if the patient makes an informed waiver of the basic informed consent information, the physician is required to honor it and may not inflict unwanted information on the patient. This should rarely happen, since patients should take their role in shared decision making seriously; and their advocates should strongly encourage patients to make important treatment decisions themselves.

What is the purpose of a consent form?

The consent of the patient is put in writing for the same reason that most contracts are in writing: to preserve the exact terms of the consent in case of future disagreement. For example, if a patient later sues a doctor or a hospital and alleges lack of informed consent, they will be able to present the written consent form in court as *evidence that consent was in fact obtained.* If the form is specific in its terms, and if it was voluntarily signed by a competent patient who understood the information set forth, the patient has very little chance of winning such a lawsuit.[26] There are, nonetheless, very few situations where a written consent form is legally required.

What is a "blanket" consent form?

A blanket consent form is one that covers (like a blanket) almost everything a doctor or a hospital might do to a patient, without mentioning anything specifically. Many hospitals continue to require patients to sign such forms on admission. A typical form will read: "I, the undersigned, hereby grant permission for the administration of any anesthetic to, and for the performance of any operation upon myself as may be deemed advisable by the surgeons in attendance at No Mercy Hospital."

What is the legal effect of signing a blanket consent form?

The general rule is that a blanket consent form is legally inadequate for any procedure that has risks or alternatives. Usually, the more vague and indefinite the terms in the consent form, the more specifically the form will be construed against the doctor or the hospital by a court.[27]

In one case, for example, a woman had consented only to a simple appendectomy, but the surgeon decided to also perform a total hysterectomy. She had signed the following consent form:

Authority to Operate

I hereby authorize the physician or physicians in charge to administer such treatment and the surgeon to have administered such anesthetics as found necessary to perform this operation which is advisable in the treatment of this patient.

The court had little difficulty deciding that this "so-called authorization is so ambiguous as to be almost completely worthless." The court determined to give it "no possible weight under the factual circumstances" of the case.[28]

Blanket forms can, however, be properly used on admission to the hospital as proof of consent to noninvasive "routine hospital procedures" (like taking blood pressure), as long as specific informed consent is obtained whenever any invasive procedure is contemplated.

Even in states having statutes that say the patient's signature on the consent form is conclusive evidence that informed consent was obtained, the signed form is only evidence that the informed consent process took place, and the patient may still present evidence that it did not.[29]

What must an "informed" consent form contain?

An "informed" consent form must contain all of the information needed to comply with the elements of informed consent that is, a description of the proposed procedure, its risks and benefits, the alternatives and their benefits and risks, the risks of nontreatment, the success rates, problems of recuperation, and the other necessary information. In general, the consent form will also contain the name(s) of the physician(s) involved. It may also include specific clauses dealing with photographs and disposition and use of removed tissues, organs, and body parts. *Patients have the right to cross out any clauses they do not agree with or consent to.*

It should be noted that although the patient may put some limits on a surgeon's authority in the written consent form, a surgeon who believes that the limitations are so strict that the surgeon cannot proceed with the operation consistent with good medical practice might be justified in declining to perform it. And the surgeon has the right to have all patient-imposed conditions noted in the record, together with the fact that the risks of such conditions have been fully explained and consented to by the patient.

What do most patients think of consent forms?

In one typical study,[30] the first question was, "What are consent forms for?" Approximately 80 percent of patients responded: "To protect the physician's rights." The authors were upset at this response, but the patients, of course, were correct. That is the primary function of *forms*. If forms are also to protect the patient, three simple steps are necessary: (1) the forms must be complete; (2) they must be in lay language; and (3) patients must be given a copy and time to think over the information contained therein before being asked to sign them. The reason none of this is usually done is that *informed consent is not usually taken seriously in the hospital setting*. It is seen as a luxury secondary to caring for the medical "needs" of the patient, and besides (it is often argued), it really does not matter anyway because patients cannot remember anything they have been told.

Another significant finding of this survey was the way patients view the consent process: Eighty percent thought the forms were necessary; 76 percent thought they contained just the right amount of information; 75 percent thought the explanation given was important; 84 percent understood all or most of the information; and 90 percent said they would try to remember the information contained on the forms. This suggests that patients understand the informed consent process very well, and that for almost all patients, informed consent is very important.

A survey done for the President's Commission for the Study of Ethical Problems in Medicine confirms the conclusion that the public strongly supports the doctrine of informed consent. Eighty-eight percent of the public believe a patient's right to information about treatment risks and alternatives should be protected by law, and only about one in five think that the time doctors spend discussing diagnosis, prognosis, and treatment could be better spent taking care of patients. A majority of the public (52 percent), as compared with 32 percent of physicians, believe that the requirements for informed consent are "clear and explicit"; and although 73 percent of physicians think the doctrine places too much emphasis on the disclosure of remote risks, less than half (44 percent) of the public feel that way. More than 40 percent of both patients and physicians believe that the amount of information disclosed should be based on what the particular patient needs to know rather than on what "a reasonable patient" needs to know, or a disclosure standard set by physicians themselves.[31]

Are clauses in consent forms in which the patient agrees not to sue the health care facility or physician enforceable?

No. A patient cannot effectively waive his or her right to sue a health care facility or physician in the event of malpractice (unless state law provides for binding arbitration upon mutual agreement). The leading case involved a patient in a nonprofit hospital. He signed a consent form that included the following clause:

RELEASE: The hospital is a nonprofit, charitable institution. In consideration of the hospital and allied services to be rendered and the rates charged therefore, *the patient* or his legal representative *agrees to and hereby releases* the Regents of the University of California [the hospital was maintained by the University], and *the hospital from any and all liability for the negligent or wrongful acts or omissions of its employees,* if the hospital has used due care in selecting its employees.[32] (emphasis added)

The main issue in the case was the validity of this clause. The Supreme Court of California found that the patient was at a great disadvantage in bargaining, compared with the hospital, and thus was almost forced to sign any agreement the hospital presented him. The court found further that this agreement affected the public interest (since the hospital held itself out as one that would perform services for any person qualified by a medical condition for service), and that requiring such a waiver as a condition of treatment was illegal and void as against public policy. The same reasoning would likely be applied to void such clauses in almost any consent form.

Can legal consent be given without signing a consent form?

Yes. As noted previously, the primary reason for a writing is to maintain a permanent record of what was agreed to. No writing is required to make most contracts, and no written form is required to make consent to treatment valid. Consent may also be implied by actions, such as the voluntary submission to treatment.

One of the first consent cases to reach the courts in the United States, for example, involved a woman who was given a smallpox vaccination. There was no explicit verbal or written consent. The court concluded that she consented to it by standing in the vaccination line, observing what was happening, and holding up her arm for the doctor.[33] The basis for allowing voluntary submission to imply consent is that the patient understood what was going on and should have been aware that her actions would be interpreted by a doctor as consenting to the procedure. For a more complex procedure, like electroshock treatment, consent by action or inaction alone is never sufficient.[34]

Treatment may also be rendered without going through the formalities of consent if the life or health of the patient is in immediate danger and obtaining consent is impossible. For example, if the victim of an automobile accident is brought in bleeding and unconscious, and no relative can be reached, treatment to save his life may be commenced immediately. While this has sometimes been justified as "implied consent," the more accurate view is that society gives doctors a "privilege" to treat patients under such extreme conditions without obtaining their consent.

Can a patient withdraw consent after signing a consent form?

Yes! *Consent* must be freely given, and so given, it *can be freely withdrawn at any time.* This is the rule, but there are practical limitations. For example, once a patient is under general anesthesia and on the operating table, it is obviously too late for the patient to change his or her mind. A written consent form in no way affects the patient's ability to change his or her mind and withdraw consent. After orally indicating one's change of mind, it may be a good idea either to obtain and destroy the original consent form or to execute another form—this time a "nonconsent form," noting on it the date and time of day consent was withdrawn.

Can a patient refuse treatment?

Yes. *A competent patient can refuse any treatment.* There are some very difficult issues involving children, which are discussed in the following chapter. There are also some unique issues dealing with potentially lifesaving or life-prolonging treatment discussed in Chapter XII, "Care of the Dying." The only other major difficult categories of patients are the mentally ill and those with communicable diseases. In general, even mentally ill patients may refuse treatment—although if they are dangerous, they may be confined to institutions for the mentally ill against their will.[35] Likewise, individuals with contagious diseases can almost always refuse treatment, but public health authorities may be able to quarantine or involuntarily hospitalize and isolate them if they are a danger to others.[36]

The reason for refusal may be as rational as the slim chances for success or as "irrational" as a fear of hypodermic needles. No matter what the reason, the refusal is just as legally binding on the doctor and the hospital. As a leading legal textbook explains:

> The very foundation of the doctrine of informed consent is every man's right to forego treatment or even cure if it entails what for him are intolerable consequences or risks, however warped or perverted his sense of values may be in the eyes of the medical profession, or even of the community, so long as any distortion falls short of what the law regards as incompetency. *Individual freedom here is guaranteed only if people are given the right to make choices which would generally be regarded as foolish.*[37] (emphasis added)

What is the hospital's duty when a patient refuses a specific treatment?

It is the obligation of the hospital, in the face of a patient's refusal, to make sure that no member of the hospital staff performs the refused procedure. If the hospital does not successfully prevent unauthorized procedures, it may be legally liable to the patient to the same extent as the doctor or other staff person who performed the procedure. The hospital is also legally obligated to continue to render the best medical care possible within the limitations imposed by the patient's refusal. Although a patient may refuse to consent to any treatment, it has been properly said by one court that the patient cannot "demand mistreatment."[38] Only if a patient consistently refuses to participate in any treatment program might the hospital, after making diligent attempts to verbally persuade the patient to change his mind, be justified in asking the patient to leave. Under these circumstances, the hospital will probably ask the patient to sign a release form that explains the proposed treatment to the patient, sets forth his refusal to consent to it, and releases the hospital from liability for the consequences of the refusal.[39] This release form is binding on the patient. For a discussion of the problem of premature discharge and discharge against the patient's wishes, *see* Chapter V, "Admission and Discharge."

How does a patient know if the patient has been properly informed?

The following checklist will help a patient determine the quality of informed consent. This list should be reviewed before the consent form is signed:

1. I know the name and nature of my injury, illness, or disability, and I know what the dangers or disadvantages of not treating it are.
2. I know the nature of the procedure which is recommended for the specific purpose of dealing with my problem.
3. I know whether or not there are other ways of treating my problem, and if there are, I have been told of the risks and benefits of these other procedures. I believe that the procedure proposed is the best one for me. I know what the advantages and benefits of this procedure are. (List the alternatives.)
4. I know what the risks, disadvantages, and side effects of this procedure are. (List these if you can.)
5. I know what the probability of success is. (What is it?)
6. I know what is likely to happen if I am not treated. (What is it?)
7. I understand all that I have been told, and I can explain the procedure in my own words. (Try to explain it to your closest friend or relative.)
8. My doctor has answered all of my questions openly and has offered to discuss any additional concerns with me. (Make sure all your questions are answered before you sign the consent form.)
9. I understand the meaning of all the words in the consent form. (If not, have them explained to you.)
10. I agree to everything in the form I signed and have crossed out things I did not agree with (or added in some new re-

quirements), and my doctor is aware of these changes. (If
you do not agree to everything in the form, do not sign it.)

11. I know the identity and qualifications of the person or persons who will be performing this procedure. (If not, find out.)

12. I have a clear head and an alert mind, and I am not so anxious
or so harassed that I feel my decision is not my own free
choice.

13. I think that the benefits I might get from this procedure are
sufficiently important to me to outweigh the risks I am
taking. (Or you should reconsider your decision.)

14. I know I do not have to consent to this procedure if I do not
want to.

When doctors and patients take informed consent seriously, the
doctor–patient relationship becomes a true partnership, with shared
authority, shared decision making, and shared responsibility for outcomes. This is a constructive and humane goal for both patients and
physicians.

The law has championed this movement toward shared decision
making through shared information but may not have gone far
enough to ensure that patients in fact get the information they need.
As explained in detail Chapter XIV, "Medical Malpractice," a patient cannot win a lawsuit against a physician for failure to disclose
the information needed for an "informed consent" unless a physical
injury is suffered (either a nondisclosed risk is actually suffered by
the patients or an undisclosed alternative treatment would have been
selected and have proven less harmful to the patient). Therefore, it
has been suggested that patients be allowed to sue physicians who
fail to disclose risks and alternatives because such a failure is an
affront to the patient's dignity as a person and denies the patient the
right to rational participation in his health care.[40] Such a legal action would be similar to suing in battery (though the injury inflicted
would be neither intentional nor physical), and the amount of money
awarded to the patient would depend on the affront to the patient's
dignity inflicted by denying the right to participate in the treatment
decision. If information sharing does not soon become the norm,
the courts will likely recognize this type of lawsuit in the future.[41]

NOTES

1. *Pratt v. Davis*, 118 Ill. App. 161, 166 (1905), *affd*, 244 Ill. 30, N.E. 562 (1906).
2. President's Commission for the Study of Ethical Problems in Medicine and Biomedical and Behavioral Research, *Making Health Care Decisions* (Washington, DC, Government Printing Office, 1982), at 2–3.
3. B. Siegel, *Love, Medicine, and Miracles* (New York: Harper & Row, 1986), at 52. Physicians may, however, seriously underestimate risks. *See* Kronlund & Phillips, *Physician Knowledge of Risks of Surgical and Invasive Diagnostic Procedures*, 142 West. J. Med. 565 (1985).
4. *See, e.g.*, Andrews, *Informed Consent Statutes and the Decisionmaking Process*, 5 J. Legal Med. 163 (1984).
5. *Schloendorff v. New York Hospital*, 211 N.Y. 127, 129, 105 N.E. 92, 93 (1914).
6. *E.g., Scott v. Wilson*, 396 S.W.2d 532 (Tex. Civ. App. 1965).
7. *See, e.g.*, J. Waltz & F. Inbau, *Medical Jurisprudence*, ch. 11, "Liability for Failure to Obtain 'Informed Consent' to Customary Therapy" (New York: Macmillan, 1971), at 152–68; Plante, *An Analysis of "Informed Consent,"* 36 Fordham L. Rev. 639 (1968); McCoid, *A Reappraisal of Liability for Unauthorized Medical Treatment*, 41 Minn. L. Rev. 381(1957); Note, *Restructuring Informed Consent: Legal Therapy for the Doctor–Patient Relationship*, 79 Yale L. J. 1533 (1970); Shultz, *From Informed Consent to Patient Choice: A New Protected Interest*, 95 Yale L. J. 219 (1985); *Natanson v. Kline*, 186 Kan. 393, 350 P.2d 1093 (1960), *reh'g denied*, 187 Kan. 186, 354 P.2d 670 (1960); *Cobbs v. Grant*, 8 Cal. 3d 229, 502 P.2d 1(1972); *Canterbury v. Spence*, 464 F.2d 772 (DC Cir. 1972), *cert. denied*, 409 U.S. 1064 (1972); *Brown v. Dahl*, 705 P.2d 781 (Wash. App. 1985); *Hook v. Rothstein*, 316 S.E.2d 690 (S.C. Ct. App. 1984).
8. E. Crouch & R. Wilson, *Risk/Benefit Analysis* (Cambridge, Mass.: Ballinger, 1982).
9. G. J. Annas, L. H. Glantz & B. F. Katz, *Informed Consent to Human Experimentation* (Cambridge, Mass.: Ballinger, 1977), at 33–38.
10. *Cobbs v. Grant*, 8 Cal. 3d. 229, 502 P.2d 1(1972). Patient dependency and physician control is also reinforced by the setting of doctor–patient interactions in which the doctor may stand, fully clothed, while the patient is lying down, often clothed only in a johnny, in a hospital bed. *See* J. Stoeckle, ed., *Encounters Between Patients and Doctors* (Cambridge, Mass.: MIT Press, 1987).
11. J. Katz, *The Silent World of Doctor and Patient* (New Haven, Conn.: Yale U. Press, 1984).

12. *E.g., Trogun v. Fruchtman,* 58 Wis. 2d 569, 207 N.W. 2d 297 (1973).
13. *E.g., Salgo v. Leland Stanford Jr. University Board of Trustees,* 154 Cal. App. 2d 560, 317 P.2d 170 (1957).
14. *E.g., Canterbury v. Spence,* Cobbs v. Grant, supra note 7.
15. *Cobbs, supra* note 7.
16. *Cf. Salgo, supra* note 13. On AIDS testing, *See* R. Bayer, *Private Acts, Social Consequences: AIDS and the Politics of Public Health* (New York: Free Press, 1989), at 101–68.
17. *Hunter v. Brown,* 4 Wash. App. 899, 484 P.2d 1162 (1971), *affd,* 81 Wash. 2d 465, r 2 r.2d 194 (1972).
18. In Sullivan v. O'Conner, 363 Mass. 579, 296 N.E.2d 183 (1973). The patient, a female entertainer, sued her physician, who she alleged had guaranteed her that her nose would be gracefully shaped following surgery. In fact, it became more grossly disfigured. She was awarded damages on the theory that the doctor had entered into a contract with her to alter her nose for the better and had guaranteed the results.
19. In *Hackworth v. Hart,* 474 S.W.2d 377 (Ky. 1971), the doctor had told a husband that a vasectomy was "a foolproof thing, 100 percent," and the court held that this stated a cause of action against the doctor when the husband's wife became pregnant. Here disclosure of the *probability of success* would seem critical, since the *only* purpose for having the procedure performed was complete sterilization. And see cases in *Annot.,* 27 A.L.R.3d 906 (1969).
20. *Demers v. Gerety,* 85 N.M. 641, 515 P.2d 645 (1973).
21. Annas & Densberger, *Competence to Refuse Medical Treatment: Autonomy v. Paternalism,* 15 Toledo L. Rev. 561, 577 (1984).
22. *Id.* at 578.
23. *Id.* at 561–78. *And see* Applebaum & Grisso, *Assessing Patients' Capacity to Consent to Treatment,* 319 New Eng. J. Med. 1635 (1988).
24. *See Petition of Nemser,* 51 Misc. 2d 616, 273 N.Y.S.2d 624 (Sup. Ct. 1966).
25. *Cobbs v. Grant,* 502 P.2d 1, 12 (1972).
26. *E.g., Karp v. Cooley,* 349 F. Supp. 827 (S.D. Texas 1972), *affd,* 493 F.2d 408 (5th Cir. 1974), *cert. denied,* 419 U.S. 845 (1974) (involving the consent form for the first human artificial heart implant).
27. *E.g., Valdez v. Percy,* 35 Cal. App. 2d 485, 96 P.2d 142 (1939); *Moore v. Webb,* 345 S.W.2d 239 (Mo. 1961).
28. *Rogers v. Lumbermens Mutual Casualty Co.,* 119 So. 2d 649, 652 (La. 1960).
29. *Parikh v. Cunningham,* 493 So. 2d 999 (Fla. 1986).
30. Cassileth et al., *Informed Consent: Why Are Its Goals Imperfectly Realized?,* 302 New Eng. J. Med. 896 (1980).
31. *Making Health Care Decisions, supra* note 2, at 105. Forms alone, of course, guarantee neither that useful information is conveyed nor that it is understood. *See* Byrne, Napier & Cuschieri, *How Informed Is Signed Consent?,*

296 Brit. Med. J. 839 (1988) (27 percent of surgical patients who had signed consent forms did not know which organ had been operated on).

32. *Tunkl v. Regents of University of California,* 60 Cal. 2d 92, 94, 32 Cal. Rptr. 33, 34, 383 P.2d 441, 442 (1963).
33. *O'Brien v. Cunard S.S. Co.,* 154 Mass. 272, 28 N.E. 266 (1891).
34. *E.g., Woods v. Brumlop,* 71 N.M. 221, 377 P.2d 520 (1962).
35. G. J. Annas, *Judging Medicine* (Clifton, NJ: Humana Press, 1988), at 23–43.
36. This is based on the broad "police powers" that states have (and can exercise through their departments of public health) to protect the health and safety of the public.
37. Harper & James, *The Law of Torts* (1968 Supp.) sec. 17.1, at 61. This remains the current rule. *See Making Health Care Decisions, supra* note 2.
38. *John F. Kennedy Hospital v. Heston,* 58 N.J. 576, 279 A.2d 670 (1971).
39. A typical release form is:

    I, [name], refuse to allow anyone to [treatment]. The risks attendant to my refusal have been fully explained to me, and I fully understand that I will in all probability need [treatment], and that if the same is not done, my chances for regaining normal health are seriously reduced, and that in all probability, my refusal for such treatment or procedure will seriously imperil my life. I hereby release the Hospital, its nurses and employees, together with all physicians in any way connected with me as a patient, from liability for respecting and following my express wishes and direction.

    The form will also contain a clause setting forth religious objections if that is the basis for the patient's refusal.
40. *See* Mariner, *Informed Consent in the Post-Modern Era,* 13 Law & Social Inquiry 385, 405 (1988); and Meisel, *A "Dignitary Tort" as a Bridge Between the Idea of Informed Consent and the Law of Informed Consent,* 16 Law, Medicine & Health Care 210 (1988).
41. Physicians recognize that informed consent can be used educationally and can positively and powerfully affect a doctor–patient alliance or partnership to their mutual benefit. As Dr. Drummond Rennie has advised his physician colleagues: Physicians must do more than just "improve consent forms" but must see "education...as a worthwhile therapeutic goal...to give patients equality in the covenant by educating them [as a counselor and advocate] to make informed decisions" (*Informed Consent by "Well-Nigh Abject" Adults,* 23 New Eng. J. Med. 917, 918 [1980]).

VII
Surgery

This chapter continues the discussion of informed consent. Although all of the principles addressed in the preceding chapter apply to surgery, surgery deserves its own chapter because of the large number of surgical procedures performed, and because surgery has such a dramatic impact on the patient, for better or worse. There are approximately thirty-five million surgical procedures done annually in the United States, about one surgical procedure for every seven people. The likelihood of any particular person undergoing a surgical procedure in any year, however, is less than this (probably about one in ten) because some people have multiple operations. Surgery accounts for most hospital admissions, more than half of all health care expenditures, and more than half of all malpractice allegations and lawsuits. Because the indications for surgery are often controversial, and usually involve questions of preferred lifestyle, this is an area in which an informed patient can be his or her own most important advocate. As one surgeon has warned, "Remember this about surgery: There are risks. There are benefits. There are choices. There are alternatives. It is your body. It is your life. The final decision is yours."[1] The "four laws of medicine" for physicians are also worth recounting:

1. If it's working, keep doing it.
2. If it's not working, stop doing it.
3. If you don't know what to do, don't do anything.
4. Never call a surgeon. (At least, never call a surgeon unless you want an operation.)[2]

What are the most common surgical procedures?

The most common surgical procedures in the United States, in rank order, are first, cesarean section (809,000 and increasing); second, hysterectomy (673,000); third, diagnostic dilation and curettage of the uterus (632,000); fourth, cataract extraction (630,000); fifth, open occlusion or destruction of the fallopian tubes (568,000);

and sixth, tonsillectomy and adenoidectomy (478,000).[3] Four of the country's six most common surgical procedures are performed exclusively on women. There are also good data to indicate that many of these procedures are medically unnecessary and are thus putting these women at unnecessarily increased risk of death and serious bodily disability.[4]

What are second-opinion programs?

Because physicians disagree about the appropriateness of surgery, and because surgery is so expensive, insurance companies and Medicare have developed "second-opinion programs." These programs either permit or require patients contemplating certain types of non-emergency surgery to get a "second opinion," from an independent surgeon, about whether the proposed surgery is appropriate. These programs were initiated to save money, and there is some evidence that they accomplish this goal by reducing the number of persons who undergo specific types of treatment. For example, in one small study of coronary artery bypass graft surgery, more than half of eighty-eight patients originally recommended for the surgery chose not to undergo it after getting a second opinion.[5] No adverse consequences were suffered by members of the group who did not undergo surgery. Critics of second-opinion programs point out that surgeons differ on when many types of surgery are indicated, and there is no way to know whether the second opinion is any better than the first.[6]

Patients have the right to get a third or fourth opinion as well, although not all insurance companies, HMOs or managed health plans pay for these additional opinions.

If surgeons cannot agree on what the best treatment is, how can a patient decide which advice to accept?

The answer to this question involves knowing who is going to be operated on and who is going to have to live with the consequences of the operation. As surgeon George Crile has put it: "There is no answer to this question, because there is no 'best' treatment. For example, are we speaking of best in terms of eradicating the disease, or best in terms of the safety of the patient, or best in terms of the comfort of the patient—or of what is best for the surgeon, in terms of safety from malpractice suits, the size of fee, or the time and trouble required to perform the operation?"[7]

As he and others have noted, there may be very important differences among alternative treatment possibilities in terms of life-style and quality of life, as well as in terms of survival. These outcomes will be valued or discounted by specific individuals on the basis of their own life-styles and value system.[3] Making a choice among treatment alternatives, in other words, is not a medical decision but a personal one that only the patient can rightly decide.

Do patients have a right to know the experience of their surgeon with a particular operation?

Yes. Knowing the experience of the surgeon and anesthesiologist is a vital piece of information needed to give informed consent. If the surgeon or anesthesiologist refuses to tell a patient what his or her experience is with the recommended procedure, the patient should get another doctor. In general, surgeons who perform operations frequently have better outcomes than those who perform the same procedure less frequently (although this is not always true). In addition, some hospitals have much poorer outcomes for specific types of conditions than others. Every year since 1985, the Department of Health and Human Services has released mortality (death) rates of approximately 6000 acute general hospitals that admit Medicare patients for specific high-risk diagnostic categories. This information is readily available and can be consulted to compare the hospitals in your area.[8] These statistics have been criticized because they fail to take severity of illness into account and thus may not fairly reflect the quality of a hospital's care. Nevertheless, although the hospital with the higher death rate may be the best one, the patient's surgeon should be asked to explain the higher death rate, and if the explanation is not satisfactory, a hospital with a lower death rate for the particular diagnosis should be considered.

As important as the experience and outcome history of the surgeon and hospital are, the experience and competence of the anesthesiologist is perhaps even more important. Some of the gravest risks of surgery are from anesthesiology. The patient should feel comfortable with the anesthesiologist (who may only show up to talk with the patient the evening prior to surgery) and the type of anesthesia that will be used. No question is too trivial, and patients should not consent to anesthesia until they are satisfied that the person in charge during the operation is competent and knows the patient's history, including prior drug reactions and allergies, completely.

Does information about alternative treatments actually affect decisions patients make regarding surgery?

Yes. In the case of breast cancer, for example, where there is continuing controversy over which treatment is most effective, more than a dozen states have passed laws specifically requiring that surgeons disclose alternative treatments to their patients. A survey of breast cancer treatment by the American College of Surgeons indicates that these laws have directly affected the treatment decisions women make. Figures for lumpectomies or wedge excisions of tumors under one centimeter were, in 1981: national average, 4.8%; Massachusetts (two-year-old law) 18%; California (one-year-old law) 10%; New York (no law) 2%.[9] New York passed its own statute in 1985, and other states are likely to continue this trend.

It makes no logical sense to enact statutes on informed consent that apply only to specific diseases, since this implies that each disease should have its own statutory disclosure requirements, rather than that the general rule requiring informed consent should simply be applied to *all* treatments. Nonetheless, it is easily understandable why breast cancer treatment has been singled out for this legislative action. Psychiatrist and expert on informed consent, Jay Katz, has noted, for example, that radical mastectomy was first proposed as breast cancer treatment by William Halsted in 1894. Halsted's proposal was based on a claim of fifty cures that has never been duplicated. His method was unchallenged in medicine until the 1950s, and it was not until the 1970s that alternatives were seriously and widely considered and studied. We have learned much more about breast cancer and have learned that "surgery, from its most limited to its most extensive varieties, as well as radiation therapy, chemotherapy, hormone therapy, and immunotherapy in various combinations and permutations of their own and with surgery have therapeutic impact...we seem to know that some treatment is better than none."

But we do not know which treatment is best for any patient. In Katz's view, this provides the physician with an opportunity and obligation to share uncertainty in treatment with the patient:

> What we do not know...is which treatment is best. Most likely any or all of the theories of cure will be modified or discarded over time. They attest to our vast ignorance about the big picture....Of importance to decision making between physician and patient is the fact that, if they are so inclined, physicians can now make clearer distinctions between what they know and do not know. Thus, they can offer patients a variety of treatment options based on pieces of evidence from available clinical data. *There is no certainty about the available knowledge, but its uncertainty can be specified.* This crucial point holds true for the treatment not only of breast cancer but for many other diseases as well.[10]

Marcia Lynch, in her poem *Peau d'Orange,* has described how difficult it is for patients confronted with a diagnosis of breast cancer not to resort to magical thinking instead of trying to weigh treatment options rationally. To her doctor she says:

> I prayed you to pull magic
> out of your black leather bag

And later promises her doctor the impossible:

> If you lift the chill
> that unravels my spine,
> I will send you stars from the
> Milky Way.
> send them spinning down,
> dancing a thousand-fold. Please
> let me grow old.[11]

The medical literature also confirms that physicians have great difficulty discussing mortality and complication rates of surgery, especially cancer surgery, with patients.[12]

What is "ghost surgery?"

Ghost surgery occurs when someone other than the surgeon the patient expects to perform the operation actually performs it. As long as the expected surgeon is present and supervising, this may not be a major issue. However, when a surgical resident or other surgeon unknown to the patient performs the procedure, and the surgeon is not physically present,[13] an unethical and illegal act is committed that includes both fraud on the part of the surgeon and battery on the part of the person who actually performs the procedure. There is no foolproof way to prevent someone other than the surgeon a patient contracts with from performing the surgery. Some precautions, however, will help. The patient should impress upon the surgeon how important it is that the surgeon personally perform the procedure, limit the written consent to the operation to the specific surgeon, and ask another physician whom the patient trusts to be present during the operation as an observer.

Can the patient designate a specific surgeon to perform surgery, or require that a specific physician be present at surgery?

Yes. Patients can condition consent to surgery (or any procedure) by consenting to it only if a specific surgeon performs the procedure, or if a specific physician, such as the patient's personal physician, is present during the procedure.[14] If surgery is nonetheless done by another physician, or in the absence of the physician whose presence the patient has required, those performing the surgery are guilty of battery, treating without consent.[15] Of course, if the patient's preconditions are unreasonable, the treatment may simply not be done. Requiring that surgery be performed without blood transfusions, for example, may or may not be reasonable, depending on the surgery involved.[16]

If a surgeon discovers an unforeseen condition during an operation, may the surgeon treat that condition if the patient has not consented to it?

When a surgeon discovers a condition that requires immediate attention, an additional procedure may be performed without consent if the condition constitutes an emergency. For example, in one case a surgeon operated on a woman he thought had a tubal pregnancy but discovered that she had acute appendicitis. A court concluded that the removal of the appendix was permissible.[17]

Even when no emergency exists, a procedure may be extended

when the extension is minor and reasonably within the scope of the consented-to procedure. In one case, a physician performing an appendectomy discovered enlarged cysts on the patient's ovaries and punctured them. The court holding this extension proper, stated: "The surgeon may extend the operation to remedy any abnormal condition in the area of the original incision whenever he, in the exercise of his sound judgment, determines that correct surgical procedure dictates and requires such an extension of the operation originally contemplated."[18]

But no extension is permissible that would result in the loss of an organ or a normal body function in the absence of an emergency. For example, a surgeon who performed an ovariectomy he claimed was necessary because of the condition of the patient's ovaries discovered during an appendectomy was found liable for this extension of the operation.[19] And neither should the scope of the operation be extended beyond the operative field originally consented to. In one case, a patient consented to an operation on her right ear; the surgeon decided that the left ear needed surgery and also operated on that ear while she was still under anesthesia. The court concluded that the operation on the left ear constituted a battery.[20]

The general theory behind the cases permitting extension of operations is that a patient consents both to having a condition treated and to assuming certain risks. As long as the procedure is extended to treat the general problem the patient sought to have treated and does not increase the risk, or cause disabilities the patient did not agree to, the extension will be deemed proper.

Of course, if a patient expressly forbids any extension of the operation, it should not be undertaken. If it can be foreseen that an extension of a procedure may be necessary before the procedure is performed, the patient's explicit consent to the extension should be sought. Patients should make their feelings on extensions known to the surgeon, and areas where extensions might occur should be specifically discussed, and decisions made about each, before surgery.

May a surgeon treat an emancipated minor without parental consent?

Yes. An emancipated minor is one who is no longer under the care, custody, and control of his parents. Thus, a married minor or one who is in the armed forces is always considered emancipated from his parents. Generally, a minor who is living apart from his parents and is self-supporting is emancipated for the purpose of consenting to medical care. Some states have codified the emancipated minor rule. In Massachusetts, for example, a minor who is living apart from his parents or guardian and is "managing" his own financial affairs may consent to medical treatment by statute, with certain exceptions.[21]

May "mature minors" consent to medical care?

No court in recent history has found any health care provider liable for treating a minor more than fifteen years old without parental consent, where the minor consented to his own care. Courts have found that a nineteen-year-old (when the age of majority was twenty-one) could consent to the use of a local anesthetic even though his mother had specifically stated she wanted a general anesthetic administered,[22] that a seventeen-year-old could consent to a vaccination required by his employer,[23] and that a seventeen-year-old could consent to the minor surgical care necessary to treat an injury incurred when her finger was caught in a door.[24] In essence, the mature minor rule means that when the minor is capable of understanding the nature, extent, and consequences of the medical treatment (information needed to give informed consent), the minor is able to consent to his or her own care.

Cases utilizing the mature minor rule generally involve older minors, treatments that are rendered for the benefit of the minor (not organ transplant or blood donors), and treatments that are necessary when some reason exists for not seeking permission of a parent, including the refusal of the minor to do so. Additionally, the treatment usually involves standard, relatively low-risk procedures.

Are there other situations in which minors may consent to their own treatment?

Yes. More than half the states have laws permitting minors to be treated for several diseases or conditions without parental consent. A significant number permit children to consent to treatment of drug dependency. A large number also permit pregnant minors to consent to medical care associated with the pregnancy (though many states exclude consent to abortion from these statutes).

As a practical matter, there is no case in which a physician who has treated a minor for pregnancy or drug dependence has been successfully sued for rendering such care, even where no statute exists permitting such treatment without parental consent. It is extremely unlikely that any health care provider who renders standard medical treatment for venereal disease, any other communicable disease, pregnancy, or drug dependency, in a situation where the minors refuse to consult their parents, would be held liable merely for rendering such care. Minors with good reasons for not wanting their parents to know about these conditions should expect physicians to treat them without parental consent.

If a parent refuses to consent to the treatment of a child, can the hospital treat the child?

There are certain cases when a hospital can treat a child although the parent refuses to consent. When a child's life is in danger and the parents will not permit treatment, courts will authorize treatment. The hospital must go to court and have someone (usually a member of its staff) appointed the guardian of the child for the purpose of consenting to the treatment. In an emergency, this can be done over the phone. If there is not even time for a phone call, emergency treatment can be rendered without any consent if needed to save the child's life or preserve the child's health.

In most such cases, the parents refuse consent on the basis of their religious convictions. In a typical case, the parents, Jehovah's Witnesses, refused to grant permission for necessary blood transfusions in the treatment of their infant son's heart abnormality. The hospital went to court and had the superintendent of the hospital appointed guardian. The superintendent then consented to the transfusion.[25]

In one unusual case, a court ordered an operation to correct a condition that did not endanger the life of the minor. A fifteen-year-old boy was suffering from Von Recklinghausen's disease, which caused a severe facial deformity. The mother, a Jehovah's Witness, would not consent to the blood transfusion necessary to the operation. The court permitted the operation to proceed without the mother's consent.[26] More commonly, the courts will not interfere in a nonlife-threatening parental decision. For example, the court would not override a father's refusal to consent to a surgical procedure necessary to repair his son's harelip and cleft palate.[27]

Who is responsible for the payment of a minor's medical bills?

One of the advantages of minority is the right to disaffirm or refuse to fulfill a contract. This means that a minor may be able to avoid the obligation to pay for services rendered by a physician or in a hospital. One exception to this rule is that minors are "liable for the reasonable value of necessaries furnished" to them.[28] Necessaries include what is reasonably needed for the minor's subsistence, including food, lodging, and education. Legally, doctor and dentist fees, and bills for medicines, are considered necessaries, so a minor is obligated to pay for them. The minor's parents might also be responsible for the cost of necessaries supplied to an unemancipated minor. But a parent is not responsible for the necessaries of emancipated minors. And for a parent to be responsible for the medical bills of an unemancipated minor, it must be shown that the parent negligently failed to provide such services although the parent knew they were necessary to the well-being of the child.

If a parent does consent, can a child refuse to undergo treatment?

There is very little law that can be relied on to answer this question. In general, however, it seems correct to conclude that a minor who is competent to consent to treatment has the right to refuse the treatment as well. In one case, a mother wanted to force her sixteen-year-old, unmarried, pregnant daughter to have an abortion over the minor's objections. A state statute gave a minor the same capacity to consent to medical treatment as an adult when the minor sought treatment concerning pregnancy. The court found that medical treatment concerning pregnancy included abortion and held:

> The minor, having the same capacity to consent as an adult, is emancipated from control of the parents with respect to medical treatment within the contemplation of the statute. We think it follows that if a minor may consent to medical treatment as an adult upon seeking treatment or advice concerning pregnancy...the one consenting has the right to forbid.[29]

May a doctor or hospital tell a child's parents about the child's condition or treatment against the child's will?

As stated in Chapter XI, "Privacy and Confidentiality," the doctor has an ethical duty (and probably a legal duty) not to disclose information received from the patient or the treatment provided to the patient. Is there anything in the parent–child relationship that should deprive the minor of confidential dealings with the doctor?

The parent is generally the consenting party, and to give informed consent, the parent needs complete information from the doctor. On the other hand, it would seem that anyone who can consent to his or her own care has a right to a confidential relationship with the doctor. In a jurisdiction in which an older minor can consent to care, the parents would not need the information for consent purposes, and the doctor should not supply it—or at best should not solicit it without telling the child in advance that whatever is learned will be disclosed to the child's parents. The purpose of confidentiality and the testimonial privilege is to encourage the patient to disclose all relevant information without worrying about embarrassing information being told to others. A child who can consent to treatment has a legal right to refuse to speak to a doctor, or to submit to explanation or treatment, until the doctor agrees not to tell the child's parents anything about the examination or treatment.

May a hospital prevent or restrict visits to children by parents?

In general, if the law or the hospital requires the parent's consent for treatment of the child, the hospital cannot prevent or restrict parents from being with their children while in the hospital.

Restrictive visiting hours for parents, where they still exist, have their roots firmly planted in a past that no longer has any relevance. When Children's Hospital of Boston was founded in 1869, for example, its purpose included bringing children "under the influence of order, purity and kindness." To isolate children from their less-than-desirable working-class home environments, visiting hours were restricted to one relative at a time, from eleven to noon on weekdays only. The intent and effect was to bar working parents from frequent contact with their children.[30] Modern hospitals have attempted to come up with other reasons to restrict visits by parents (for example, it makes children who do not have parents who can stay with them all the time uncomfortable), however, none have any more validity than the 1869 visiting restrictions.

But it was not until the 1950s that some in the medical profession began to view as paradoxical the fact that "when a young child needs his mother most, when the child is ill and perhaps in pain, she is generally not allowed to be with the child for more than brief visits."[31] Since then, moves have continued toward liberalizing visiting hours in pediatric wards and hospitals, and parents who insist can stay with their child in almost every modern hospital.

In Boston, a consumer group called Children in Hospitals was organized in the early 1970s to support parents who want to stay with their children in the hospital and to encourage hospitals to broaden their visiting policies.[32] It was founded by a mother who demanded and was granted the right to stay with her ten-month-old daughter during a hospitalization. At one point, a nurse put some drops in the child's eye and left. The baby started screaming and would not open her eyes. Twice the mother carried the child out to the nursing station and was given no satisfaction. After a third try, she persuaded the nurse to check to be sure the right drops had been administered. The nurse discovered that the iodine drops were for the baby to drink prior to a brain-wave test. Her eyes were immediately flushed out, and luckily no permanent damage was done.

The legal right to be with one's children derives from the doctrine of informed consent. Parents may not be able to give fully informed consent for their children if they are not able to be with them constantly to monitor their reactions (which they can interpret better than anyone else because of their experience with their children). Also, parents have the right to withdraw their consent to treatment at any time, and this right can only be meaningfully exercised if the parent is continuously present with the child to determine that circumstances have changed to such an extent that consent should be withdrawn. Parents whose request to stay with their child is refused can limit any consent they are asked to give, or form they are asked to sign, with the condition that they be permitted to stay with their child. If they are thereafter denied the right, they can terminate their consent, and the hospital can no longer legally treat their child.

The only reasonable limits a hospital can probably place on parental visitation would involve actual interference with the hospital's ability to care for other patients (not the parent's own child, since the parents and not the doctor or hospital have the ultimate treatment authority). This would mean that parents, if they so desire, have the legal right to stay with their children during all tests and procedures, the induction of anesthesia, and to be present in the recovery room when the child regains consciousness.

What can parents do to make hospitals more responsive to their desires?

Question hospital policy before your child is admitted.

Select the doctor and hospital best able to make the arrangements you desire.

Negotiate directly with the chief of pediatrics or the chief of anesthesia, never with the admissions office or floor nurses.

Publicize your experiences, both favorable and unfavorable.

Form parent groups to work for the changes you want.

Can assertiveness and attempts to be independent hurt a patient's recovery?

The contrary seems to be true: Assertive patients are those involved in their care and who want to do all they can to get better. Surgeon Bernie Siegel, who asserts that "the most important kind of assertiveness a patient can demonstrate is in the formation of a participatory relationship with the doctor," suggests eleven steps, which he calls "Good Patient, Bad Patient," that he encourages patients who are going to the hospital to follow. Some are not legally related, but patients have the legal right to do all of them:

1. For your hospital stay, take clothes that are practical, comfortable, and individual. Plan to walk as much as possible.
2. Take room decorations of a personal and inspirational nature.
3. Question authority—tests, etc. Speak up for yourself, your needs and comfort in all areas, both before and during tests.

4. Make your doctor aware of your unique needs and desires...
5. Take tape recorder and earphones....Record conversations with your physician for later review and for family use.
6. Use your tape recorder in the operating room and recovery room to hear music, meditation, or messages during and after surgery. Have someone put a reminder into the doctor's orders to play the entire tape continuously.
7. ...Instruct the surgeon and anesthesiologist to repeat positive messages to you. The simplest is that you will awaken comfortable, thirsty, and hungry.
8. Tell the surgeon to speak to you during surgery, honestly but hopefully...
9. Speak to your own body, particularly the night before surgery, suggesting that the blood leave the area of surgery and that you'll heal rapidly.
10. Arrange visits and calls from those who will nurture and love you, as well as give you "carefrontation" when appropriate.
11. Get moving as soon as possible after surgery. Leave the hospital to attend group meetings, go for walks, or have meals out with friends.[33]

Not all patients will want to follow these suggestions or are the "exceptional patients" that Siegel encourages all of us to be. But the point should not be lost: Patients have a legal right to do these things if they so choose. And if the surgeon does not permit such things as the patient listening to music during surgery, the patient has a right to fire the surgeon and hire someone who will respect the patient's wishes. Remember: "It is your body. It is your life. The final decision is yours."[34]

NOTES

1. G. Crile, *Surgery* (New York: Delacorte Press, 1978), at xvi.
2. M. Konner, *Becoming a Doctor* (New York: Penguin Books, 1988), at 21.

3. Rutkow, *General Surgical Operations in the United States,* 121 Arch. Surg.
    1145 (1986).
4. *See generally* Boston Women's Health Book Collective, *The New Our Bod-
    ies, Ourselves* (New York: Simon & Schuster, 1984).
5. Graboys et al., *Results of a Second-Opinion Program for Coronary Artery
    Bypass Graft Surgery,* 258 JAMA 1611(1987); *and see generally* Fried-
    man, "Second-Opinion Programs Come into Their Own," *Hospitals,* July
    16, 1984, at 105.
6. *E.g.,* Clarke, *A Comparison of Decision Analysis and Second Opinion for
    Surgical Decisions,* 120 Arch. Surg. 844 (1985).
7. *Supra* note 1, at xiii.
8. *E.g.,* in Consumer Checkbook, *Consumers' Guide to Hospitals* (Washing-
    ton, DC: Washington Consumers' Checkbook, 1988). (Send $8 for a copy
    of this book to Washington Consumer's Checkbook, 806 15th St. NW, Suite
    925, Washington, DC 20005 (202) 347-9612. *But see* Tolchin, "Data on
    Hospital Mortality Rates Are Challenged," New York Times, July 11, 1988,
    at A13; and Dubois *et al., Hospital Inpatient Mortality: Is It a Predictor of
    Quality?* 317 New Eng. J. Med. 1674 (1987).
9. Dabice & Cordes, "Informed Consent Heralds Change in Breast Treatment,"
    Medical News, Nov. 11, 1985, at 1, 4.
10. J. Katz, *The Silent World of Doctor and Patient* (New Haven, Conn .: Yale
    U. Press, 1984), at 183–84. A *US News and World Report* cover story on
    breast cancer concluded: "The fact is that since so much of breast cancer
    treatment is experimental, no one knows the bottom line of what benefits
    whom. Three well-qualified doctors may suggest three completely differ-
    ent treatment plans. So no woman diagnosed with breast cancer can afford
    to let her doctor do all the deciding." "How to Beat Breast Cancer," July 11,
    1988, at 52, 56. The good news is that modern cancer specialists "welcome
    a woman's active participation in treatment decisions" (Siegel, "Deciding
    Factors: New Options in Breast Cancer Therapy Can Save a Woman's Life
    and Her Sense of Self," *Self,* Feb. 1989, at 127). For information on cancer
    treatment and research call the National Cancer Institute (NCI) (800-
    4CANCER).
11. The entire poem appears in J. Mukand, ed., *Sutured Words: Contemporary
    Poems about Medicine* (Brookline, Mass.: Aviva Press, 1987), at 124.
12. *See, e.g.,* Mayer & Patterson, *How Is Cancer Treatment Chosen?* 318 New
    Eng. J. Med. 636 (1988).
13. *See, e.g.,* Rensberger, "The Touchy Ethical Issue of Trainee Surgeons Op-
    erating without Consent of Patients," New York Times, Feb. 6, 1978, at
    B12.
14. *Johnson v. McMurray,* 461 So. 2d 775 (Ala. 1984).
15. *Pugsley v. Privette,* 220 Va. 892, 263 S.E.2d 69 (Va. 1980).
16. For information on surgeons who will honor a patient's request not to do

blood transfusions, call 1-800-NO-BLOOD. On the question of whether the blood of surgeons should be screened to see if they are infected with HIV, *see* Gostin, *HIV-Infected Physicians and the Practice of Seriously Invasive Procedures,* 19 Hastings Center Report 32 (Jan. 1989).

17. *Barnett v. Bachrach,* 34 A.2d 626 (Mun. Ct. App. DC 1943).
18. *Kennedy v. Parrott,* 243 N.C. 355, 362, 90 S.E.2d 754, 759 (1956).
19. *Wells v. Van Nort,* 100 Ohio St. 101, 125 N.E. 910 (1919).
20. *Mohr v. Williams,* 95 Minn. 261, 104 N.W. 12 (1905).
21. Mass. Gen. L. ch. 112, sec. 12(F).
22. *Bishop v. Shurly,* 237 Mich. 76, 211 N.W. 75 (1926).
23. *Gulf & S.I.R. Co. v. Sullivan,* 155 Miss. 1, 119 So. 501 (1928).
24. *Younts v. St. Francis Hospital,* 205 Kan. 292, 469 P.2d 330 (1970).
25. *State v. Perricone,* 37 N.J. 463, 181 A.2d 751(1962).
26. *In re Sampson,* 65 Misc. 2d 658, 317 N.Y.S.2d 641(1970), *affd* 29 N.Y.2d 900, 278 N.E.2d 918 (1972).
27. *In re Seiferth,* 309 N.Y. 80, 127 N.E.2d 820 (1955).
28. Simpson, *Contracts,* sec. 105.
29. *In re Smith,* 16 Md. App. 209, 225, 295 A.2d 238, 246 (1972).
30. P. Starr, *The Social Transformation of American Medicine* (New York: Basic Books, 1982), at 158.
31. Robertson, *Young Children in Hospitals* (New York: Basic Books: 1958); *and see* Diesenhouse, "Suffering of Children Is Eased as Hospitals Change to Keep Families Near," *New York Times,* Dec. 15,1988, at B22.
32. For further information write Barbara Popper, Children in Hospitals, 31 Wilshire Park, Needham, Mass. 02191 (617) 482-2915.
33. B. Siegel, *Love, Medicine and Miracles* (New York: Harper & Row, 1986), at 173–74.
34. *Supra* note 1.

VIII
Pregnancy and Birth

The law has taken the lead in delineating the contours of the public debate over reproductive liberty. Most of that debate, in turn, has focused on the Supreme Court's landmark 1973 decision in *Roe v. Wade*.[1] This decision followed two other major opinions on "the right to privacy" and has itself been followed by more than a dozen other decisions on abortion. A few words about these important decisions provide an introduction to this chapter.

The first case involved the constitutionality of a Connecticut statute that made it a crime to use contraceptives. In striking down this statute, the Supreme Court enunciated a constitutional "right to privacy" suggested by "penumbras" that emanate from specific guarantees in the Bill of Rights. Defining a specific "zone of privacy," Justice William Douglas, writing for the Court, focused on sexual relations in marriage:

> We deal with a right of privacy older than the Bill of Rights—older than our political parties, older than our school system. Marriage is the coming together for better or for worse, hopefully enduring, and intimate to the degree of being sacred. It is an association that promotes a way of life, not causes; a harmony in living, not political faiths; a bilateral loyalty, not commercial or social projects...an association for as noble a purpose as any involved in prior decisions.[2]

To secure the argument, the Court also noted that the statute could not be enforced without massive and unthinkable governmental intrusion into people's lives and homes: "Would we allow the police to search the sacred precincts of marital bedrooms for telltale signs of the use of contraceptives? The very idea is repulsive to the notions of privacy surrounding the marriage relationship."

Strictly speaking, the Connecticut case applied only to married couples. But during the same term in which *Roe v. Wade* was being discussed, the Court decided another case involving a Massachusetts statute that made it a crime to sell, lend, give away, or exhibit any contraceptive device. In that case, the Court concluded that there was no constitutionally acceptable rationale for treating married and unmarried individuals differently with respect to contraception, and that the basis for the right to privacy was the sexual relationship itself, not marriage:

> The marital couple is not an independent entity with a mind and heart of its own, but *an association of two individuals* each with a separate intellectual and emotional makeup. If the right to privacy means anything, it is the right of the *individual*, married or single, to be free from unwarranted governmental intrusion into matters so fundamentally affecting a person as the decision whether to bear or beget a child.[3] (emphasis added)

This important decision presaged *Roe* by going further than the Court had to go in broadening the right to privacy to include a decision not only to "beget" a child but to "bear" one as well.

The 1973 opinion of *Roe v. Wade* is the most important, most controversial, and most well known US Supreme Court decision in recent history. At issue in *Roe* was a Texas statute that made it a crime to "procure an abortion" or to attempt one, except to save the life of the mother. Justice Harry Blackmun wrote the opinion of the Court, which determined that the "right to privacy" existed "in the Fourteenth Amendment's concept of personal liberty." The Court concluded that this right "is broad enough to encompass a woman's decision whether or not to terminate her pregnancy": "The detriment that the state would impose upon a pregnant woman by denying this choice altogether is apparent. Specific and direct harm medically diagnosable even in early pregnancy may be involved. Maternity, or additional offspring, may force upon a woman a distressful life and future."[4]

The Court also recognized, however, that the state has interests in the health of the mother and the life of the fetus that may at times be sufficiently "compelling" to permit the state to limit abortion. With regard to maternal health, the Court found this interest compelling only after the point in pregnancy at which abortion becomes more dangerous to the woman's health than carrying the fetus to term (in 1973, about the end of the first trimester). With regard to fetal life, the Court found the state could legitimately claim a compelling

interest in its protection after viability, which the Court defined as
the time the fetus was capable of surviving independently of its mother
(still about the beginning of the third trimester). The Court thus
concluded that the state could regulate abortion procedures to protect
the pregnant woman's health after the time in pregnancy when
abortion was no longer safer than childbirth, and that it could outlaw
pregnancy terminations or elective premature deliveries altogether
(except when the life or health of the pregnant woman was at stake)
after the fetus was viable.[5]

In dividing pregnancy into three stages and weighing state interests in regulating abortion differently during each, the US Supreme
Court was acting more like a legislature than a court. The decision
has been severely criticized on this basis, but it has survived nearly
two decades of intense attack and remains the law of the land. In
1989, in the *Webster* decision, the Supreme Court narrowly affirmed
Roe v. Wade, but all indications are that future Courts will chip away
at that decision.[5] Whatever the Court does, however, it and the legislatures of the individual states will be faced with the same question:
What is the appropriate role for the government to play in regulating
the decisions of women regarding pregnancy termination?

Is the consent of a woman's husband ever necessary for any type of treatment of a married woman?

No. The consent of a husband is never legally necessary for the
treatment of a conscious, competent, and consenting married woman.
This has been consistently affirmed by the courts. Specific cases
considered in the context of a husband suing the doctor for treating
his wife without his consent have included both sterilization and
pregnancy care.[6] In each, the husband's consent was determined not
to be necessary. This finding, of course, works both ways; the wife's
consent is not required for any type of medical treatment of her husband.

The US Supreme Court has ruled that states cannot enact legislation that permits husbands to have veto power over their wives' decisions to have an abortion, since, among other things, "it is the woman
who bears the child and is more directly and immediately affected
by the pregnancy."[7] An Oklahoma court has ruled, consistent with
other courts, that husbands also have no right to prevent their wives

from becoming sterilized. In the court's words: "We have found no authority and the plaintiff has cited none which holds that the husband has a right to a childbearing wife as an incident of their marriage. We are neither prepared to create a right in the husband to have a fertile wife nor to allow recovery for damage to such a right. We find the right of the person who is capable of competent consent to control his own body paramount."[8]

Hospitals and doctors who use consent forms with spaces requiring the consent of the patient's husband should abolish these anachronistic relics of sexism and update their forms to reflect twentieth-century law, medicine, and sexual equality. Patients asked to sign such a form should simply cross out the "signature of husband" line.

Does a patient being examined by a doctor of the opposite sex have a right to have another person of the patient's sex present in the examining room during the examination?

Yes. The overwhelming majority of specialists in obstetrics and gynecology are men. Most male physicians routinely have a female nurse with them during pelvic examinations. The reason generally given is to protect themselves from a possible charge by the woman patient of improper advances. If the doctor's practice is not to have another female present, however, a woman does have the right to demand that another female be present during the examination if she desires. If the doctor refuses this request in the hospital context, his method of practice is open to serious ethical question, and his conduct should be reported immediately to the chief of his service. The name and phone number of his chief can be obtained from one of the nurses or through the hospital administration. In a private office or clinic, the woman's remedy is probably limited to walking out, informing the local medical society and state licensing agency, and going to another doctor.

Feminists often advise women to "bring a friend" with them on visits to the doctor for psychological support. Such a person not only can act as a witness to what goes on but can also remind the patient to ask certain questions that are bothering her, and make sure instructions and recommendations regarding treatment are understood. No ethical physician should object to this request.

Does a woman have a right to refuse to be examined by medical students, interns, or residents in a hospital setting?

All patients have a right to refuse to be examined by *anyone* in the hospital setting. When a woman is asked, "Do you mind if these other doctors look at you also?" she has every right to say, "Yes, I do mind," and refuse to permit them to examine her. One should be especially wary of the phrase "young doctor," since this almost always means medical student—usually a first- or second-year one. In some hospitals, medical students are also referred to simply as "doctors." Nonetheless, the patient has a right to know both the extent of their training and the purpose of their proposed examination.

In one court case from the 1930s, a woman objected to being examined by a medical student just before she was due to give birth. Thereupon, an older doctor came in, performed a rectal and vaginal examination, and then had the same examination performed two or three times each by "ten or twelve young men who she took to be students." She protested repeatedly and testified, "Whenever I screamed and protested they just laughed, told me to shut up." She experienced both emotional and physical damages from the delivery. The court had no difficulty in finding this conduct "revolting" and an assault on the patient. In the court's words, "A physician or a medical student has no more right to needlessly and rudely lay hands upon a patient against her will than has a layman." The court also determined that the hospital in which this event took place could be held liable for permitting "unlicensed students to experiment on the patient and treat her without her consent."[9]

The lesson is that not only the medical students themselves but also the attending physician and the hospital are liable to the patient for any unauthorized examination or treatment. Consent by a patient, based on a belief that the person examining her is a doctor, is consent achieved by fraud or misrepresentation (if the examiner is not a doctor) and, as such, is not legally valid.

Do pregnant woman have a right to genetic counseling and screening?

Yes. All pregnant women should be advised of the existence of genetic tests for the fetus and should be given the option of availing themselves of these tests. Physicians caring for pregnant women have an obligation to inform them of the existence and uses of such tests as chorionic villus sampling (CVS), maternal serum alphafetoprotein screening (MSAFP), ultrasound, and amniocentesis. The American College of Obstetricians and Gynecologists (ACOG) Standards state:

> Antenatal screening for genetic disorders is an integral part of obstetric care....The history obtained during the initial evaluation should be reviewed to detect signs of possible genetic disorders.... The following factors are indicative of such risk:
> Advanced maternal age (35 years or older at expected time of delivery)
> Previous offspring with a chromosomal aberration, particularly autosomal trisomy
> Chromosomal abnormality in either parent
> Family history of a sex-linked condition
> Inborn errors of metabolism
> Neural tube defects
> Hemoglobinopathies.
> Amniocentesis, fetoscopy, chorionic villus sampling, ultrasound examination, and cytogenetic assessment are some of the tests used for antenatal genetic diagnosis. Physicians should not hesitate to seek consultations with specialists in antenatal diagnosis of genetic disorders [and the patient should not hesitate to ask for a referral if one is not offered].[10]

Do physicians and patients have a right to use "alternative" forms of child delivery in hospitals?

Yes, although this is more a matter of moral and political right than legal right. In fact, courts have been very reluctant to interfere with hospital delivery room policy. This issue has been most directly addressed by "husband in the delivery room" cases. Most hospitals with maternity wards permit husband-coached childbirth (such as Lamaze or psychoprophylactic) in which the father is with the mother throughout labor and delivery. Nonetheless, lawsuits to compel a change in the policies of those institutions that do not allow this have so far been unsuccessful. This is true whether the hospital is private or public, and whether the plaintiff is a pregnant woman or her obstetrician.

In the first appellate decision on this question, a Montana court found that a hospital rule that forbade the presence of fathers in the delivery room was not arbitrary or capricious, the court refused to intrude itself "into the administration of the hospital where the hospital had acted in good faith on competent medical advice."[11]

Another case involved a suit by prospective parents and their physician against a public hospital. They contended that the hospital's policy denying fathers access to the delivery room violated constitutional rights, specifically, the "right of marital privacy" enunciated in the birth control and abortion decisions. The lower court dismissed their suit, and the appeals court affirmed in a two to one decision.[12]

Where does this leave the hospital-bound couple? In general, the couple and the physician they choose must follow hospital policy in regard to deliveries (including such things as the presence of the father and others and the method of delivery). The woman, however, does have the right to refuse any particular medical procedure or drug offered (for example, fetal monitor, anesthesia, episiotomy, HIV screening). Control is primarily reactive, since the patient cannot demand that things happen in a certain way but can only refuse certain things that are offered. Given the proper hospital environment and supportive health care professionals, this may be entirely satisfactory. If it is not, the only alternative is for women to locate a more responsive hospital or to give birth outside of the hospital, either in a birth center or at home.

There are two simultaneous and conflicting trends in modern obstetrics. The first is toward a more relaxed and informal birthing process in the hospital. The second is toward higher and higher cesarean section rates (25 percent in the United States in 1988) in the United States. The first trend is the result of demands by pregnant women and increasing competition for their business. The second is much more complicated and seems to involve reimbursement, possibly liability concerns, a belief that at times it may be best for the neonate, the use of fetal monitors, and indications for cesarean sections that are unique to the United States.[13] Almost everyone agrees that our cesarean section rate is too high, but no one seems to know how to slow down the ever-increasing rate.[14]

Does a woman have a right to refuse a cesarean section?

Yes. Competent adults have the right to refuse any medical intervention, and women do not lose this right just because they are pregnant or in labor. The fetus (soon-to-be child) has interests (that may even be called rights) as well, but no matter how highly we regard the fetus, we cannot justify forcing invasive and possibly dangerous surgery on its mother against her will anymore than we can force her to submit to surgery that might help her child. Forcing surgery degrades pregnant women and dehumanizes them; treating them like inert containers. There has been a highly charged and emotional debate on this subject, and at least twenty-one "forced cesarean sections" have been performed in the United States under court order.[15] These courts, however, acted in emergency settings, under tremendous pressure, in cases generally involving poor women from other cultures with whom the physicians could not identify. The law provides no more support for these decisions than it would for one that would force a mother to "donate" a kidney or bone marrow to her child.[16]

It will be the rare woman who can withstand the arguments of an obstetrician recommending a cesarean section when she is in active labor; but refusal is her right. Even the American College of Obstetricians and Gynecologists recognized this right in a 1987 policy statement exhorting its members to talk to the woman in labor, rather than to a judge, when conflicts arise. This is good advice.

Judges are terrible at making emergency medical decisions, and it is inappropriate for judges to act impulsively, without benefit of reflection on past precedent and the likely future impact of their legal opinions.[17] The cesarean section cases all suffer from lack of reflection. The delivery room is not conducive to such reflection, and judges do not belong there at all in such circumstances.[18]

The law must honor the rare case of a woman's refusal. This may seem callous to the rights of fetuses, since some fetuses that might be salvaged may die or be born severely handicapped. This will be tragic, but it is likely to be rare. It is the price society pays for protecting the rights of all competent adults and preventing forcible, physical violations of women by coercive obstetricians and judges. The choice between fetal health and maternal liberty is laced with moral and ethical dilemmas, but the force of law and the intervention of the courts and police will not make them go away.

Does a dying pregnant woman have the right to refuse a cesarean section (or other surgical intervention designed to save the fetus) if the fetus is viable?

Yes. The fact that a pregnant woman is terminally ill or dying does not strip her of her constitutional rights; pregnant women have no obligation to undergo invasive and potentially life-threatening surgical procedures. Nonetheless, in 1987, a young pregnant woman was forced to undergo a cesarean section in what can probably best be described as a medical and legal atrocity: the case of "Angela C." The case merits detailed discussion in this book in the hope that knowledge of it will be sufficient to prevent a similar thing from ever happening to another pregnant woman.[19]

Angela C was a 26-year-old married woman who had suffered from cancer since she was 13. About 25 weeks into her pregnancy, she was admitted to George Washington University Hospital, and a massive tumor was found in her lung. Physicians determined that she would die within a short time. Her husband, her mother, and her physicians agreed that keeping her comfortable while she died was what she wanted, and that her wishes should be honored. This was communicated to a hospital administrator, who called legal counsel, who in turn invited a judge to the hospital to decide what to do.

A District of Columbia Superior Court judge rushed to the hospital and set up "court." After a hearing, the judge issued his opinion orally. The centerpiece was Ms. C's terminal condition. In the judge's words, "The uncontroverted medical testimony is that Angela will probably die within the next 24–48 hours." He did "not clearly know what Angela's present views are" about the cesarean section, but found that the fetus had a 50–60 percent chance to survive and less than a 20 percent chance for serious handicap. The judge concluded: "It's not an easy decision to make, but given the choices, the Court is of the view the fetus should be given the opportunity to live."

The court reconvened shortly thereafter when informed that Ms. C, who had been unconscious, was awake and communicating. The chief of obstetrics reported that she "clearly communicated" and "very clearly mouthed words several times, 'I don't want it done. I don't want it done. I don't want it done.' " Nonetheless, without even talking to Ms. C herself, the judge reaffirmed his original order. Less than an hour later, three judges heard a request for a stay, over the telephone, and denied it.

The cesarean section was performed, and the nonviable fetus died approximately two hours later. Angela C, now confronted with both recovery from major surgery and the knowledge of her child's death, died approximately two days later. Five months later, the Court of Appeals issued its written opinion.[20] The opinion reads more like a sympathy card than a judicial pronouncement. Its first paragraph, for example, ends with the following sentence: "Condolences are extended to those who lost the mother and child." The opinion is fatally flawed. The most serious error is the statement that "as a matter of law, the right of a woman to an abortion is different and distinct from her obligations to the fetus once she has decided not to timely terminate her pregnancy." This is incorrect as both a factual and a legal matter. Ms. C never "decided not to timely terminate her pregnancy," and because of her fetus' effect on her health, under *Roe v. Wade* she could have authorized her pregnancy to be terminated (to protect her health) at any time prior to her death. Moreover, had the roles been reversed, and an abortion was required to save Ms. C's life, no legal principle would permit a judge to order the abortion against her will.

Another basis on which the opinion rests is that a parent cannot refuse treatment necessary to save the life of a child (true), and therefore a pregnant woman cannot refuse treatment necessary to save the life of her fetus (false). The child must be treated because parents have obligations to act in the "best interests" of their children (as defined by child neglect laws), and treatment in no way compromises the bodily integrity of the parents. Fetuses, however, are not independent persons and cannot be treated without invading the mother's body. Treating the fetus against the will of the mother requires us to degrade and dehumanize the mother and treat her as an inert container. This *is* acceptable once the mother is dead, but it is never acceptable when the mother is alive. The court seems to understand this and thus ultimately justifies its opinion on the basis that Ms. C was as good as dead, and had no "good health" to be "sacrificed." "The cesarean section would not significantly affect A.C.'s condition because she had, at best, two days of sedated life."

But this reasoning will not do. It would, for example, permit the involuntary removal of vital organs prior to death when they were needed to "save a life." But if the child had already been born, no court would require its mother to undergo major surgery for its sake (for example, a kidney "donation") no matter how dire the potential consequences of refusal to the child.

This unprincipled opinion was thankfully vacated by the full bench in early 1988 and specifically overruled in 1990 when the court essentially adopted the ACOG/medical standard, and could think of no "extremely rare and truly exceptional" case in which the state might have an interest sufficiently compelling to override the pregnant woman's wishes.[21] Nevertheless, it dramatically illustrates the general rule that judges should never go to the hospital to make emergency treatment decisions. Rushed to an unfamiliar environment, asked to make a decision under great stress, and having no time either for reflection or for study of existing law and precedents, a judge cannot act judiciously. Facts cannot be properly developed, and the law cannot be accurately determined or fairly applied to the facts. The "emergency hearing" scenario is an invitation to arbitrariness and the exercise of raw force.

This was *not* a hard case. If there really were facts in dispute, a case conference involving the patient, family, and all attending health care personnel could have been held to assess them. Direct communication with the patient is almost always the most useful and constructive response to "problems" like those presented here. Calling a judge is usually a counterproductive panic reaction. Patients, their families, and advocates should insist that the patient's wishes, not those of hospital lawyers or uninvolved physicians, be determined and honored.

Can pregnant women be charged with "fetal neglect" if they engage in behavior that injures their fetus?

Not under current law. But there are people who would like to write new laws, like child abuse and neglect laws, that would apply to fetuses. The problems with this approach are well illustrated by the case of Pamela Stewart, the subject of what may have been the first criminal charge ever brought against a mother for acts and omissions during pregnancy.[22] Criminal charges were filed against her in California in October 1986. Sometime very late in her pregnancy, she was reportedly advised by her physician not to take amphetamines, to stay off her feet, to avoid sexual intercourse, and, because of a placenta previa, to seek immediate medical treatment if she began to hemorrhage. According to the police, she noticed some bleeding the morning of November 23, 1985. Nevertheless, she allegedly remained at home, took some amphetamines, and had intercourse with her husband. She began bleeding more heavily, and contractions began sometime during the afternoon. It was only later, perhaps "many hours" later, that she went to the hospital. Her son was born that evening, had massive brain damage, and died about six weeks later.

The case against her was dismissed in early 1987 on the grounds that the California child support statutes did not apply to this conduct. Should the law be changed to create a new crime of fetal neglect? Does it make any sense to decree that a pregnant woman must, in effect, live for her fetus? That she must legally "stay off her feet" if walking or working might induce contractions? That she commits a crime if she does not eat only healthy foods; smokes cigarettes or drinks alcohol; takes any drugs (legal or illegal); becomes infected with HIV; has intercourse with her husband? Should all these "dos" and "don'ts" be catalogued in a statute, or should they be the subject of her physician's advice? And how would such a criminal law change the nature of the doctor–patient relationship?

It seems evident that in the minds of the police the doctor's patient was not Ms. Stewart but her fetus. It also seems evident that although called "advice" and "instructions," the police believed that the physician was actually giving Ms. Stewart orders—orders that she *must* follow or face criminal penalties, including jail. This is nonsensical and dangerous: nonsensical because medical advice should remain advice—physicians are neither lawmakers nor seers; dangerous because medical advice is a vague term that can cover almost anything.

After-the-fact prosecutions would also not help individual fetuses. Effectively monitoring compliance would require actual confinement of pregnant women to an environment in which eating, exercise, drug use, and sexual intercourse could be controlled. This could, of course, be a maximum-security country club, but such massive invasions of privacy can only be justified by treating pregnant women as nonpersons during their pregnancy.

Other quandaries must be faced to apply child neglect statutes to fetuses. The first is that, unlike a child, the fetus is absolutely dependent on its mother and cannot itself be "treated" without in some way invading the mother. Favoring the fetus radically devalues the pregnant woman and treats her like an inert incubator or as a culture medium for the fetus.

Child neglect covers a wide variety of activities, but generally involves failure to provide necessities, like clothing, food, housing, or medical attention to the child. Such laws do not, however, require parents to provide optimal clothing, food, housing, or medical attention to their children, and they do not even forbid parents taking risks with them (such as permitting them to engage in dangerous sports) or affirmatively injuring them (such as corporal punishment to teach them a lesson). None forbids mothers to smoke, take dangerous drugs, or to consume excessive amounts of alcohol, even though these activities may have a negative effect on their children.

If society really wants to help pregnant women protect the health of their fetuses, the answer is not to criminalize dangerous behavior during pregnancy, but to provide access to supportive prenatal care, education, and counseling. Until such services are universally available, criminalization of behavior does not merit serious discussion.[23]

May a woman legally plan to give birth at home?

Yes. There are no laws in any state that require childbirth to take place in a hospital. The parents' primary duty is to the potential child. No state requires women to take affirmative action to safeguard their fetuses. All states, however, have child abuse statutes that forbid parents from abusing their children, and require them to provide their children with necessary medical attention. Parents may usually make a decision to have a home birth with impunity. If they have reason to know that complications are likely to develop that will require hospital care to save the child from death or permanent injury, however, and the child dies or is permanently injured because of the home birth, the parents could be held criminally liable. The charge would be child abuse in both cases, and possibly manslaughter (depending on the cause of death and its predictability) if the child dies.[24] This general rule should dissuade parents from attempting to manage home births by themselves and encourage them to seek a licensed attendant at the home birth or to seek hospitalization when it is indicated to protect the health of the child.

Who may attend a home birth?

In every state, physicians may attend home births. The role other attendants, such as nurse midwives and lay midwives, may play is governed by the law of the particular state.[25]

Statutes and qualifications vary, but in states that have specific legislation providing for the licensing of midwives, it is likely that courts would find anyone not a physician or licensed midwife who held himself or herself out as an expert on childbirth and attended a childbirth would be guilty of the crime of practicing either medicine or midwifery without a license (in a number of states a birth attendant must have been compensated to be prosecuted). The court's reasoning would probably parallel that of a California Supreme Court decision that concluded the legislature had an "interest in regulating the qualifications of those who hold themselves out as childbirth attendants...for many women must necessarily rely on those with qualifications which they cannot personally verify."[26] The rationale is that the legislature can reasonably decide that the only way to protect the health of the public in the matter of childbirth is to ensure that those who hold themselves out as experts to the public are required to meet certain standards and be licensed. After this conclusion is reached by the enactment of a licensing statute, anyone practicing outside its limitations would be guilty of the crime of practicing without a license.

Is it malpractice for a physician to participate in a home birth?

No. Physician fear of potentially increased malpractice liability for participation in home births is probably based on ignorance of the law of medical malpractice, combined with the climate of fear that has been created in many states. So long as the decision for a home birth is made by the woman after she has been fully informed of the risks and potential complications, and so long as all generally accepted medical steps have been taken concerning screening the woman and emergency backup facilities, it is highly unlikely that any malpractice action against a physician would be successful.

In attending childbirth and counseling the pregnant woman, the physician also has a duty toward the fetus to provide it with adequate medical care both before and during the delivery.[27] This duty does not usually prevent a physician from participating in a home delivery, but it does require the physician to perform any standard screening tests for high-risk pregnancies and to attempt to persuade pregnant women at high risk to use hospital facilities for birth.

Is there a right to procreate (to have a child)?

There is a legal right *not* to procreate, by having access to sterilization, contraception, and abortion technology through a physician. There are strong arguments against any government interference with the decision of a married couple to have children. Nonetheless, because children are involved, the government may limit some options involving the "new reproductive technologies," such as in vitro fertilization (IVF), and embryo transfer (ET). For example, the government could constitutionally enact a statute that designated the woman who gave birth to a child its legal mother, even if the egg used was from another woman (and had been implanted in her artificially), and she had previously agreed to give the child up upon birth to the "egg donor." The rationale would be that the gestational or birth mother had contributed more to the child than the egg donor, had taken a much greater physical risk, was more psychologically bonded to the child, was more easily identifiable than the egg donor, and would definitely be with the child at birth to protect it and make treatment decisions for it. Likewise, the state could outlaw the sale of human embryos even if this might make it more difficult for couples to obtain embryos for use in having a child. Other than enacting laws protecting the birth mother by making her the legal mother and outlawing the sale of human embryos (to prevent commercialization and commodification of children), the state should limit its role in human reproduction to education and the provision of reasonable services.[28]

What does ACOG say about patient rights?

The Standards of the American College of Obstetricians and Gynecologists (ACOG) provide strong support for patient rights. In particular:

Informed Consent: It is the physician's responsibility to inform the patient of the nature of the surgical or medical procedure being recommended. In most cases, the explanation should include the necessity of the treatment, the management alternatives, the reasonably foreseeable risks and hazards involved, the chances of recovery, and the likelihood of desired outcome. Adequate opportunity should be provided to encourage and answer questions, including those regarding nonmedical issues such as cost....*Patients' Rights and Responsibilities:* Recognition of the rights of patients is an important aspect of health care. Patients are entitled to the following rights:

> Impartial access to high-quality care
> Respect, dignity, and privacy
> Assurance of confidentiality of their disclosure and their records
> Knowledge of the identity and professional status of individuals providing their health care.

Patients should be advised of their diagnosis, treatment, and prognosis, and given an opportunity for informed participation in their health care. Patients have the right to request consultation or to refuse treatment. In addition, they have the right to have an appropriate representative exercise these rights when they are unable to do so for themselves.[29]

What is "The Pregnant Patient's Bill of Rights"?

This document was prepared by Doris Haire, the chair of the Health Law and Regulation Committee of the International Childbirth Education Association.[30] Although this document can be viewed primarily as a political one aimed at helping to encourage change in obstetric practices and at empowering pregnant women, almost all its provisions have at least some support in the law. Items 1, 7, 8, 9, 11, 13, and 14 can properly be labeled "legal rights," and most of the remainder, including 15 and 16, can be labeled "probable legal rights."

The Pregnant Patient's Bill of Rights

1. *The Pregnant Patient has the right,* prior to the administration of any drug or procedure, to be informed by the health professional caring for her of any potential direct or indirect effects, risks, or hazards to herself or her unborn or newborn infant which may result from the use of a drug or procedure prescribed for or administered to her during pregnancy, labor, birth, or lactation.

2. *The Pregnant Patient has the right,* prior to the proposed therapy, to be informed, not only of the benefits, risks, and hazards of the proposed therapy but also of known alternative therapies, such as available childbirth education classes that could help to prepare the Pregnant Patient physically and mentally to cope with the discomfort or stress of pregnancy and the experience of childbirth, thereby reducing or eliminating her need for drugs and obstetric intervention. She should be offered such information early in her pregnancy in order that she may make a reasoned decision.

3. *The Pregnant Patient has the right,* prior to the administration of any drug, to be informed by the health professional who is prescribing or administering the drug to her that any drug which she receives during pregnancy, labor, and birth, no matter how or when the drug is taken or administered, may adversely affect her unborn baby, directly or indirectly, and that there is no drug or chemical that has been proven safe for the unborn child.

4. *The Pregnant Patient has the right,* if cesarean birth is anticipated, to be informed prior to the administration of any drug, and preferably prior to her hospitalization, that minimizing her and, in turn, her baby's intake of nonessential pre-operative medicine will benefit her baby.

5. *The Pregnant Patient has the right,* prior to the administration of a drug or procedure, to be informed of the areas of uncertainty if there is no properly controlled followup research which has established the safety of the drug or procedure with regard to its direct and/or indirect

effects on the physiological, mental, and neurological development of the child exposed, via the mother, to the drug or procedure during pregnancy, labor, birth, or lactation—(this would apply to virtually all drugs and the vast majority of obstetric procedures).

6. *The Pregnant Patient has the right,* prior to the administration of any drug, to be informed of the brand name and generic name of the drug in order that she may advise the health professional of any past adverse reaction to the drug.

7. *The Pregnant Patient has the right* to determine for herself, without pressure from her attendant, whether she will accept the risks inherent in the proposed therapy or refuse a drug or procedure.

8. *The Pregnant Patient has the right* to know the name and qualifications of the individual administering a medication or procedure to her during labor or birth.

9. *The Pregnant Patient has the right* to be informed, prior to the administration of any procedure, whether that procedure is being administered to her for her or her baby's benefit (medically indicated) or as an elective procedure (for convenience, teaching purposes, or research).

10. *The Pregnant Patient has the right* to be accompanied during the stress of labor and birth by someone she cares for, and to whom she looks for emotional comfort and encouragement.

11. *The Pregnant Patient has the right* after appropriate medical consultation to choose a position for labor and for birth which is least stressful to her baby and to herself.

12. *The Obstetric Patient has the right* to have her baby cared for at her bedside if her baby is normal, and to feed her baby according to her baby's needs rather than according to the hospital regimen.

13. *The Obstetric Patient has the right* to be informed in writing of the name of the person who actually delivered her baby and the professional qualifications of that person. This information should also be on the birth certificate.

14. *The Obstetric Patient has the right* to be informed if there is any known or indicated aspect of her or her baby's care or condition which may cause her or her baby later difficulty or problems.

15. *The Obstetric Patient has the right* to have her baby's hospital medical records complete, accurate, and legible and to have their records, including Nurses' Notes, retained by the hospital until the child reaches at least the age of majority, or, alternatively, to have the records offered to her before they are destroyed.

16. *The Obstetric Patient,* both during and after her hospital stay, *has the right* to have access to her complete hospital medical records, including Nurses' Notes, and to receive a copy upon payment of a reasonable fee and without incurring the expense of retaining an attorney.

The Bill of Rights concludes by appropriately noting, "It is the obstetric patient and her baby, not the health professional, who must sustain any trauma or injury resulting from the use of a drug or obstetric procedure. The observation of the rights listed above will not only permit the obstetric patient to participate in the decisions involving her and her baby's health care, but will help to protect the health professional and the hospital against litigation arising from resentment or misunderstanding on the part of the mother."

NOTES

1. 410 U.S. 113 (1973). *See generally* Symposium Issue: *Justice Harry A. Blackmun: The Supreme Court and the Limits of Medical Privacy,* 13 Am. J. Law & Med. 153–525 (1987).
2. *Griswold v. Connecticut,* 381 U.S. 479 (1965).
3. *Eisenstadt v. Baird,* 405 U.S. 438, 453 (1972).
4. The US Supreme Court has further defined viability as follows:
   Viability is reached when, in the judgment of the attending physician on the particular facts of the case before him, there is a reasonable likelihood of the fetus' sustained survival outside the womb, with or without artificial support. Because this point may differ with each pregnancy, neither the

legislature nor the courts may proclaim one of the elements entering into the ascertained of viability—be it weeks of gestation or fetal weight or any other single factor—as the determinant of when the State has a compelling interest in the life or health of the fetus. Viability is the critical point. And we have recognized no attempt to stretch the point of viability one way or the other. (*Colautti v. Franklin,* 439 US 379, 388–89 [1979].)

If and when an early abortion pill or "contragestive" drug is approved for use in this country, early abortion will seem more like birth control, and the decision to use it will be almost exclusively in the hands of the woman herself. *See,* e.g., Greenhouse, "A New Pill: A Fierce Battle," New York Times Magazine, Feb. 12, 1989, at 23.

5. *Webster v. Reproductive Health Services,* 109 A. Xr. 3040 (1989); Annas, *The Supreme Court, Privacy, and Abortion,* 321 New Eng. J. Med. 1200 (1989).

6. *E.g., Kritzer v. Citron,* 101 Cal. App. 2d 33, 224 P.2d 808 (1950); *Rosenberg v. Feigin,* 119 Cal. App. 2d 783, 260 P.2d 143 (1953); *Rytkonen v. Lojacono,* 269 Mich. 270, 257 N.W. 703 (1934).

7. *Planned Parenthood of Central Missouri v. Danforth,* 428 US 52 (1976).

8. *Murray v. Vandevander,* 522 P.2d 302, 304 (Okla. Ct. App. 1974).

9. *Inderbitzen v. Lane Hospital,* 124 Cal. App. 462, 12 P.2d 744 (1932).

10. ACOG, *Standards for Obstetric-Gynecologic Services,* 6th ed. (Washington, DC: ACOG, 1985), at 18–19.

11. *Hulit v. St. Vincent's Hospital,* 164 Mont. 168, 520 P.2d 99 (1974).

12. *Fitzgerald v. Porter Memorial Hospital,* 523 F.2d 716 (7th Cir. 1975).

13. Shiono, Fielden, McNellis, *et al., Recent Trends in Cesarean Birth and Trial of Labor Rates in the United States,* 257 JAMA 494 (1987). On fetal monitors, *see* Friedman, *The Obstetrician's Dilemma: How Much Fetal Monitoring and Cesarean Sections Is Enough?,* 315 New Eng. J. Med. 641 (1986).

14. Friedman, "Consumer Group Calls Half of Nation's Cesareans Unnecessary," *Medical World News,* Dec. 28, 1987, at 70–71. In Oct. 1988, ACOG issued new guidelines to require its members to encourage women to attempt vaginal delivery, even if the woman has had one or more cesarean sections (Knox, "Inhibit Cesareans, Doctors Told," Boston Globe, Oct. 26, 1988, at 1). *See also American Medical News,* Feb. 10, 1989, at 12. C/SEC, a consumer education and support group, called the new ACOG policy a "ray of hope" (22 Forest Rd., Framingham, Mass. 01701).

15. Kolder, Gallagher & Parsons, *Court-Ordered Obstetrical Interventions,* 316 New Eng. J. Med. 1192 (1987); *and see Jefferson v. Griffin Spalding Cty. Hospital Auth.,* 247 Ga. 86, 274 S.E.2d 457 (1981).

16. Annas, Protecting the Liberty of Pregnant Patients, 316 New Eng. J. Med. 1213 (1987); *and see McFall v. Shimp,* 10 Pa. D. & C.3d 90 (Allegheny Cty. 1978) (bone marrow donation cannot be compelled by law even to save a life of a relative).

17. *Application of the President and Directors of Georgetown College,* 331 F. 2d 1000 (DC Cir. 1964).
18. Elias & Annas, *supra* note 4, at 256–60; *and see* Gallager, *Prenatal Invasions and Interventions: What's Wrong with Fetal Rights?,* 10 Harv. Women's L. J. 9 (1987). *But see* Robertson, *The Right to Procreate and In Utero Fetal Therapy,* 3 J. Legal Med. 333 (1982).
19. The facts of this case are taken from the transcript. *And see* Annas, *She's Going to Die: The Case of Angela C.,* 18(1) Hastings Center Report 23 (Feb. 1988); letters, 18(3) Hastings Center Report 40–42 (June 1988); and Burt, *Uncertainty and Medical Authority,* 16 Law, Medicine & Health Care 190, 192–95 (1988).
20. *In re A.C.,* 533 A.2d 611 (App. D.C. 1987).
21. *In re A.C.,* 573 A.2d 1235 (App. D.C. 1990) discussed in Annas, *Foreclosing the Use of Force: A.C. Reversed,* Hastings Center Rpt. 27–29.
22. Elias & Annas, *supra* note 4, at 261–62. A more detailed account is in G. J. Annas, *Judging Medicine* (Clifton, N. J.: Humana Press, 1988), at 91–96.
23. Elias & Annas, *supra* note 4, at 262. *And see* Johnsen, *The Creation of Fetal Rights: Conflicts with Women's Constitutional Rights of Liberty, Privacy and Equal Protection,* 95 Yale L. Rev. 599 (1986). The crack epidemic simply increases the urgency of making prenatal care available to the poor. *See* Mariner, Glantz & Annas, *Pregnancy, Drugs, and the Perils of Prosecution,* 9 Criminal Justice Ethics 30–41 (1990).
24. *See* Robertson, Involuntary Euthanasia of Defective Newborns, 27 Stan. L. Rev. 213 (1975).
25. *See generally* S. Sagov *et al.,* eds., *Home Birth* (Rockville, Md.: Aspen, 1984).
26. *Bowland v. Municipal Hospital of Santa Cruz,* 134 Cal. Rptr. 630, 638 (1976).
27. Robertson, *supra* note 18; and *Commonwealth v. Edelin,* 359 N.E.2d 4 (Mass. 1976).
28. Elias & Annas, *supra* note 4, at 222–42. *See also* US Congress, Office of Technology Assessment, *Infertility: Medical and Social Choices* (Washington, DC: Government Printing Office, 1988) (OTA-BA-358); Symposium Issue: *Surrogate Motherhood: Politics and Privacy,* 16 Law, Medicine & Health Care 1–137 (1988).
29. ACOG, *supra* note 10, at 84.
30. ICEA Publication Center, PO Box 9316, Midtown Plaza, Rochester, NY 14604.

IX
Human Experimentation and Research

The history of medical progress is to a large extent the history of medical experimentation. Few wish to halt the practice of experimentation on patients altogether; nonetheless, there is a growing awareness that individual rights have often been trampled in the process, and that steps must be taken to protect patients from overzealous researchers. Many of the most blatant abuses of human beings for the purpose of medical experimentation have taken place in prisons, mental institutions, residences for the retarded, and the military. All of these are important. This chapter concentrates on the major site of human experimentation: the hospital.

It is often asserted that being an experimental subject is the price one pays for being a patient in a teaching hospital. This is nonsense. People do not automatically forfeit their human rights and become transformed into laboratory animals by seeking care in a teaching hospital. But it is difficult to resist the request of a physician to be an experimental or research subject in the "total institution" environment of the hospital, where the full weight of the medical profession can be brought to bear on the patient.

Because of a history of abuse, various regulatory mechanisms have been developed to protect the rights and welfare of potential subjects. To protect that welfare, legitimate experimentation requires a sound hypothesis based on prior work (usually on animals), reason to expect beneficial results (for society, if not for the patient), a competent researcher, and prior review by a qualified panel. To protect the subject's rights, legitimate experimentation requires the voluntary, competent, informed, and understanding consent of the subject. This chapter will explore the issues at stake when people are asked to participate in research. The operative word is "asked," no patient is ever under any obligation to participate in research, and no patient is obliged to give any reason for refusing to participate.

What is medical experimentation?

No single definition of medical experimentation is entirely satisfactory. Some doctors have argued, for example, that every time any patient is treated, a therapeutic experiment is taking place, since that particular patient has never been treated at that particular time in that particular way. This view is patently incorrect, since it denies that there can ever be such a thing as "standard medical treatment" and seems to treat quacks and physicians who follow routine medical procedures identically, both always ignorant of the possible results of their treatment. It is most descriptive to think of the development from experimentation to standard medical procedure as a continuum, with "innovative therapy" or "nonvalidated therapy" somewhere in between these two types of interventions. Perhaps the most distinguishing characteristic of experimentation is the extent of uncertainty as to outcome, recognizing that some uncertainty applies to all medical procedures and treatment. For the purposes of this chapter, a physician is engaged in experimentation on a patient–subject when the physician *departs from standard medical practice* for the *purpose of obtaining new, generalizable knowledge* or to test a hypothesis, using the scientific method.

What is the most important question in medical experimentation?

The most important question is whether the experiment should be done at all.[1] Only after this determination has been made, based on such factors as prior research, risk/benefit analysis, the importance of the research to society, and the available alternatives for the patient–subject, is it legitimate to even ask the subject to participate. As will be emphasized throughout this chapter, *informed consent is a necessary, but not sufficient, condition for legitimate human experimentation.* A careful review of the science comes first.

A reasonable summary of some of the major issues in human experimentation appears in Gustave Flaubert's realistic novel *Madame Bovary* (1857). Charles Bovary decides to try to make his name as a physician by curing the local stableman's clubfoot with experimental surgery that involves screwing the foot and leg into "a kind of box, weighing about eight pounds, constructed by the carpenter and the locksmith, with a prodigal amount of iron, wood, sheet-iron, leather, nails and screws." He explains the proposal to his wife, Emma: "'What risk is there? Look!'—and he counted the 'pros' on his fingers. 'Success, practically certain. An end of suffering and disfigurement for the patient. Immediate fame for the operator.' "

The stableman was urged to consent by the entire town, but the "decisive factor was that it *wouldn't cost him anything*" (emphasis in original). The experiment did not go as planned and another physician eventually had to be called in to ampute the hideously painful and gangrenous leg. Not all experiments have such disastrous results for patients; but many share the same motivations on the part of both physician and patient, the same inability to separate wishful fantasy from a realistic appraisal of likely outcomes, and the same inability to distinguish voluntary consent from coercion.

What is the Nuremberg Code?

The Nuremberg Code is the most comprehensive and authoritative legal statement on human experimentation. This ten-point code was articulated in a 1947 court opinion following the trial of Nazi physicians for "war crimes and crimes again humanity" committed during World War II, which include experiments designed to determine which poisons killed the fastest, how long people could live exposed to ice water or when exposed to high altitudes, and if surgically severed limbs could be reattached. The court rejected the defendants' contention that their experiments with both prisoners of war and civilians were consistent with the ethics of the medical profession as evidenced by previously published United States, French, and British experiments on venereal disease, plague, and malaria and United States prison experiments, among others. The court concluded that only "certain types of medical experiments on human beings, when *kept within reasonably well defined bounds,* conform to the ethics of the medical profession generally."

The code's basis is a type of natural law reasoning. In the court's words: "All agree...that certain basic principles must be observed in order to satisfy moral, ethical, and legal concepts."[2] Principle 1 enunciates the primacy the law places on the consent of the subject:

The voluntary consent of the human subject is absolutely essential.

This means that the person involved should have *legal capacity* to *give consent*; should be so situated as to be able to exercise *free power of choice*, without the intervention of any element of force, fraud, deceit, duress, overreaching, or other ulterior form of constraint or coercion; and should have *suffcient knowledge* and *comprehension* of the elements of the subject matter involved as to enable him to make an understanding and enlightened decision. This latter element requires that before the acceptance of an affirmative decision by the experimental subject there should be made known to him the nature, duration and purpose of the experiment; the method and means by which it is to be conducted; all inconveniences and hazards reasonably to be expected; and the effects upon his health or person which may possibly come from his participation in the experiment.

The duty and responsibility for ascertaining the quality of the consent rests upon each individual who initiates, directs or engages in the experiment. It is a personal duty and responsibility which may not be delegated to another with impunity. (emphasis added)

The Nuremberg Code thus requires that the consent of the experimental subject have at least four characteristics: it must be competent, voluntary, informed, and comprehending. This is to protect the subject's rights.

The other nine principles deal primarily with protecting the subject's welfare: They prescribe actions that must be taken prior to and during the experiment. These include a determination that the experiment is properly designed to yield fruitful results "unprocurable by other methods"; that "anticipated results" will justify performance of the experiment; that all "unnecessary physical and mental suffering and injury" are avoided; that there is no "*a priori* reason to believe that death or disabling injury will occur"; that the project has "humanitarian importance" that outweighs the degree of risk; that "adequate preparation" is taken to "protect the experimental subject against even the remote possibilities of injury, disability, or death"; that only "scientifically qualified" persons conduct the experiment; that the subject can terminate participation at any time; and that the experimenter is prepared to terminate the experiment if "continuation is likely to result in injury, disability, or death to the experimental subject."

The code has earned worldwide acceptance and been used as part of the basis for other international documents, such as the Declaration of Helsinki. It is a part of international common law, and I have previously argued that it can properly be viewed as both a criminal and a civil basis for liability in the United States.[3] Unfortunately, when the United States Supreme Court had a chance to adopt and endorse the principles of the Nuremberg Code in 1987, it failed to recognize the code as binding on the United States military by a five-to-four vote.[4]

The case involved a lawsuit brought by an army serviceman, who, while in the service, had been secretly given LSD in 1958 to determine its effects. He suffered from hallucinations, periods of incoherence, and memory loss and would wake in the middle of the night and violently beat his wife and children. He was discharged from the Army in 1969 and divorced shortly thereafter because of the LSD-induced personality changes. In 1975, the Army sent him a letter asking him to cooperate in a follow-up study on the long-term effects of LSD on "volunteers who participated" in the 1958 tests. This was the first he learned of the experiment. When his suit got to the United States Supreme Court, the majority concluded that involuntary participation in human experimentation was no exception to the rule that a serviceman could not sue the United States for injuries that "arise out of or are in the course of activity incident to service."

The four dissenting judges based their disapproval of this conclusion primarily on the principles enunciated in the Nuremberg Code. Justice William Brennan, writing for three of them, concluded: "The United States military developed the Code, which applies to all citizens—soldiers as well as civilians." Justice Brennan went on to note that in addition to the thousands of soldiers and civilians who have been subjected to secret LSD experiments by the government, an estimated 250,000 military personnel were exposed to large doses of radiation while engaged in maneuvers designed to determine the effectiveness of combat troops in nuclear battlefield conditions between 1945 and 1963. Explaining why the Nuremberg Code and the principle of voluntary consent it stands for must apply to the military as well as civilians, Justice Brennan noted: "The subject of experimentation who has not volunteered is treated as an object, a sample.... Soldiers ought not be asked to defend a Constitution indifferent to their essential human dignity." This decision is a national disgrace. The Nuremberg Code should continue to set a *minimal* legal standard for licit human experimentation, both in and out of the military.

Have US physicians and hospitals been involved in illegal and unethical experiments since World War II?

Yes. The United States has had many medical experimentation scandals.[5] In a 1963 study, terminally ill patients at the Jewish Chronic Disease Hospital in Brooklyn, New York, were injected with live cancer cells to test their immune response. The patients were not informed of the type of cells injected, only that it was a "skin test."[6] In the "Tuskegee study," which was begun prior to World War II and continued for twenty-five years thereafter, effective treatment was withheld from a group of poor black rural males who had syphilis so that the natural course of the disease could be studied.[7] In the "Willowbrook experiments," retarded children were deliberately infected with hepatitis so that experimentation could continue on a potential vaccine. In the early 1960s, more than 2 million thalidomide tablets were distributed to 20,000 patients in the United States by 1267 physicians as part of a research study. When told of the potentially toxic effects of the drug on fetuses, more than four hundred physicians made no effort to contact their patients directly, and many of these were unable to do so because they had kept no records of those to whom they had given the drug.[8]

In 1966, Dr. Henry Beecher of Harvard Medical School reported on twenty-two unethical experiments that took place in the United States after the promulgation of the Nuremberg Code.[9] Examples included the deliberate withholding of penicillin from sufferers of streptococcal infection, and the "successful" transplantation of a cancerous tumor from a daughter to her mother. Congressman Ed Markey of Massachusetts has documented dozens of secret radiation experiments on unconsenting patients sponsored by the Atomic Energy Commission.[10]

Some of the most publicized experiments in history were also of questionable ethics. Most notable are experiments, like the Baby Fae case, involving animal organs transplanted into humans (xenografts).[11] Much of the experimentation involving use of the artificial heart has also suffered from serious ethical and legal problems in both research design and consent procedures.[12] For example, Barney Clark, the recipient of the world's first permanent artificial heart, signed an eleven-page consent form that is notable more for its length than for its content. It was incomplete, internally inconsistent, and confusing. It assumed, as his physicians then believed, that Clark would either die on the operating table or return home in about ten days and continue to be mentally competent for the rest of his life. It took no account at all of a "halfway success"—survival coupled with severe confusion, mental incompetence, or coma. The consent form made no provisions for proxy consent to additional procedures or experiments in the event of incompetence, for a mechanism to terminate the experiment, or for how Clark would die.[13]

These examples are spectacular and unusual; nonetheless, they demonstrate the point Flaubert made in *Madame Bovary* more than a century ago: When great fame and possible fortune are at stake, ethics and law take a back seat to even fanciful notions of "scientific advance." They also illustrate the fascination the public has with glitzy experimentation and the difficulty regulators have in dealing with ambitious physicians.

Who is likely to be experimented on in the hospital setting?

Many doctors argue that all patients in teaching hospitals should be considered appropriate candidates for experimentation. But what empirical data we have suggest that the poor are more likely to be experimented on than the middle class or the rich. One study of a major teaching hospital concluded that although private patients may not be much better off, "the ignorant, the poor, and the ethnically despised are more likely to be used as subjects, partly just because they are more often ward and clinic patients, but partly also because their handicaps make them more available."[14] Nor are they likely to be informed of what is going on. In a specific case, for example, only one of three clinic patients in a new drug study for the induction of labor in childbirth knew they were experimental subjects. Also, instead of being told an experimental drug was being used on them, they had been informed that a "new" (implying "better") drug was being used.[15]

A leading medical commentator has argued that the hospital patient is the most "at risk" for experimentation, and that the doctrine of informed consent cannot protect patients adequately. In his words:

> Volunteers for experiments will usually be influenced by hopes of obtaining better grades, earlier parole, more substantial egos, or just mundane cash. These pressures, however, are but fractional shadows of those enclosing the patient–subject. *Incapacitated and hospitalized because of illness, frightened by strange and impersonal routines, and fearful for his health and perhaps life, he is far from exercising a free power of choice* when the person to whom he anchors all his hopes asks, "say, you wouldn't mind, would you, if you joined some of the other patients on this floor and helped us to carry out some very important research we are doing?" When "informed consent" is obtained, it is not the student, the destitute bum, or the prisoner to whom, by virtue of his condition, the thumb screws of coercion are most relentlessly applied; it is *the most used and useful of all experimental subjects, the patient with disease.*[16] (emphasis added)

How is human experimentation monitored in the hospital?

As a result of publicity generated by many abuses of subjects in human experimentation and research, most hospitals now have Institutional Review Boards (IRBs) whose job is to review proposed research *before* patients are asked to enroll in it and to approve written consent forms.[17] IRBs are *required* in all hospitals that receive federal funding and in all hospitals that have filed a "general assurance" with the federal government agreeing to have the research done in their facilities so approved. This prior review is critical, since consent of the patient cannot make an unethical experiment ethical; and no patient should be asked to consent to take part in an experiment that has not been adequately reviewed. On the other hand, just because an IRB has approved the research does not mean that patients have any obligation to participate in it. Indeed, one primary purpose of the consent form is to make sure that patients understand that they have no obligation to consent to participate in any experiment or research, and that their care will not be compromised in any way if they refuse to participate.

What does an Institutional Review Board (IRB) do?

As set forth in federal regulations, an IRB's primary responsibility is to protect human subjects by reviewing proposed research and approving it if satisfied that the following requirements are met:

1. Risks to subjects are minimized.
2. Risks to subjects are reasonable in relation to anticipated benefits, if any, to subjects, and the importance of the knowledge that may reasonably be expected to result.
3. Selection of subjects is equitable.
4. Informed consent is obtained and appropriately documented.[18]

Some studies may require special procedures to monitor data to ensure the safety of subjects (such as drug studies), and to protect the confidentiality of records and may include additional provisions to safeguard subjects who are especially vulnerable to coercion or undue influence. IRBs are a classic form of regulation: the use of public incentives to induce self-regulation on the part of powerful institutions. The IRB itself must be composed of "not less than five" members of varying backgrounds. Not all can be members of the same profession or sex or associated with the same institution. In addition to professional competence in research, "the committee must be able to ascertain the acceptability of proposals in terms of institutional commitments, and regulations, applicable laws, standards of professional conduct and practice."[19]

These requirements seem stringent, but they could be minimally met by having four researchers from the institution on the IRB joined by a lawyer from the community. Since a quorum can be defined as a majority of members, the lawyer need never attend any meeting, or can be outvoted at all of them, and the "peer-review" IRB will still meet these requirements. IRBs were mandated as a reaction to the types of abuses in research outlined elsewhere in this chapter. Nonetheless, it is fair to conclude that they have done at least as much to protect researchers from more stringent controls that might otherwise be mandated as they have done to enhance the protection of subjects. *If a patient has any question about research* they (or someone else) have been asked to consent to, it is usually useful to *contact the chairman of the IRB directly.*

What are the purposes of obtaining the informed consent of research subjects?

The purposes of informed consent to human experimentation are (1) to promote individual autonomy and (2) to encourage rational decision making.[20] The purpose of autonomy (or self-determination) is to protect the individual's integrity as a person by denying anyone the right to invade the person's body without consent. This proposition can be restated in a number of ways and approached from a variety of directions—the ultimate conclusion, however, remains the same. For example, it can be thought of as the "right to be left alone," or the "right to privacy" concerning decisions about one's body. Another view is that research on human subjects must always be a "joint enterprise" between the researcher and subject to prevent the subject from becoming an object or thing to be used, instead of a human being.[21] From this perspective, informed consent is necessary to prevent taking advantage of subjects to the point where the types of procedures performed on them are expanded beyond what society would view as tolerable.[22]

Still another approach begins by noting that although the purpose of human experimentation is ultimately to benefit society through medical progress, the burdens and risks of such experimentation will of necessity fall on only a few individuals. In this view, the subject is giving a "gift" to society—a gift that cannot rightfully be forcibly taken, and one that is devalued (and the giver debased) if the gift is based on a false assumption of the risks involved.[23] Adequate information and free choice are essential to protect an individual's autonomy and personhood.

The counterargument, that respect for individual autonomy sometimes delays scientific advance, is insightfully disposed of by philosopher Hans Jonas:

> Let us not forget that *progress is an optional goal,* not an unconditional commitment, and that its tempo in particular, compulsive as

it may become, has nothing sacred about it. Let us also remember that a slower progress in the conquest of disease would not threaten society, grievous as it is to those who have to deplore that their particular disease be not yet conquered, but that *society would indeed be threatened by the erosion of those moral values whose loss, probably caused by too ruthless a pursuit of scientific progress, would make its most dazzling triumphs not worth having.*[24] (emphasis added)

Rational decision making is clearly viewed as a secondary goal by the courts. Nonetheless, it is an extremely important function of the informed consent doctrine, since if this goal is not achieved, the entire research enterprise is illegitimate.[25] After a candid review of extreme risks, subjects are likely to refuse to participate in experiments that are inherently too dangerous to be ethically performed on humans, making them *de facto* impossible to do. And this is as it should be. As Professors Jay Katz and Alexander Capron point out, "Who other than the patient–subjects can determine whether the benefits of the procedure, conventional or experimental, outweigh the burdens that will be imposed on them?"[26]

Researcher–physicians might also have conflicting motives, such as career advancement or monetary gain, and these should also be disclosed to the patient.

What are the components of informed consent to human experimentation?

Federal regulations specify that potential subjects must be given all of the following information on any proposed experiment:

1. A statement that the study involves research, an explanation of its purposes and the expected duration of the subject's participation, and a description of the procedures to be followed, identifying which are experimental
2. A description of reasonably foreseeable risks or discomforts to the subject
3. A description of any reasonably expected benefits to the subject or to others
4. A disclosure of appropriate alternative procedures or courses of treatment that might be advantageous to the subject

5. A statement describing the extent to which confidentiality of records identifying the subject will be maintained
6. For research involving more than minimal risk, an explanation about any compensation or medical treatments available if injury occurs
7. An explanation of whom to contact for answers to questions about research and research subject's rights or in the event of injury
8. A statement that participation is voluntary and that refusal to participate will involve no penalty or loss of benefits; and that the subject may discontinue at any time without prejudice.[27]

An Institutional Review Board (IRB) may waive the requirement for a signed consent form if (1) the only record linking the subject to the research would be the consent form, and this would put the subject at risk regarding confidentiality; and (2) the research presents no more than minimal risk of harm and involves no procedures for which written consent is normally required outside the research context.[28] In addition, some categories of research, such as normal educational research, educational testing, most interview procedures, some observational research, and use of public data, may be exempt from IRB review. Other categories of research, such as collection of hair and nail clippings, collection of external secretions and small blood samples, voice recordings, and moderate exercise, can be given an informal, quick review, called "expedited review," and may not require written informed consent.[29]

Can researchers delegate the task of obtaining informed consent?

Yes. A nurse or resident may ask for consent, but the Nuremberg Code makes it clear that obtaining the informed consent of the research subject is the obligation of the researcher: "The duty and responsibility for ascertaining the quality of the consent rests upon each individual who initiates, directs, or engages in the experiment." It is a *"personal duty and responsibility* which may not be delegated to another with impunity" (emphasis added).

Since this issue is not specifically addressed in the federal regulations, the only way to assure that the goals of promoting autonomy and rational decision making are met is to insist that those conducting the research are the ones who actually obtain the subjects' informed consent. Accordingly, *patients should not sign consent forms for research discussed with them only by nurses and other allied health care professionals who are not directly conducting the experiment* in question. Consent should be given only to a member of the research team who is able to answer all the questions and concerns that the patient and the patient's family have.

Is the informed consent of the research subject always necessary for lawful experimentation?

Yes. There are two critical lessons to be learned from this chapter: Informed consent is a *necessary precondition* to lawful human experimentation, but informed consent alone is not a *sufficient precondition*. Other requirements, such as the reasonableness of the experiment, review by an institutional review board, and provisions for withdrawal and compensation for harm must be met before consent is even sought.

Can the next of kin or a legal guardian consent to research to be performed on his relative or ward?

The only times such "proxy" consent is proper is if the proposed research is in the "best interests" of the patient (in the sense that the possible benefits to the patient exceed the possible risks); or if he or she has, while still competent, indicated that he or she would like to participate in this particular research after becoming incompetent, and the conditions anticipated by the now-incompetent patient actually exist; or in unusual cases, where there is minimal risk, but the knowledge to be gained is important to the group of patients to which the subject belongs and cannot be obtained any other way. When a person is asked to provide such consent, the person should always remember that there is no obligation to consent, and the person should only consent if the experiment will benefit the relative or ward because its burdens to the patient are outweighed by its potential benefits to the patient.

Incompetent patients are often used as subjects simply because they are readily available, but convenience alone is never a sufficient justification to use someone as a research subject. Competent patients are always preferable as subjects because they can consent on their own behalf.

To protect especially vulnerable populations, such as nursing home patients, it has been suggested that in addition to other protections, "a nursing home council, composed primarily of residents, should review and approve any protocol before the research can be conducted at the facility."[30] This is because what may seem trivial to the researcher and the IRB members in terms of risk, discomfort, disorientation, or dehumanizing effects may not seem so trivial to this vulnerable and often-frightened population. Proxies must always remember that their primary role is to protect the patient.[31]

Should indemnification for injury to subjects in human experimentation be made mandatory?

Yes. There are many compelling reasons why the researcher, the institution in which the research is conducted, the sponsor of the research, or the federal government should be required to provide research subjects with an insurance policy to compensate them for injuries suffered as subjects. This suggestion was made in early 1973 by the secretary of HHS's Commission on Medical Malpractice.[32] Others have also made it. Henry K. Beecher, for example, has argued: "Even if all reasonable precautions have been taken to protect both the subject and the investigator from physical damage as well as from unethical practices, the possibility of injury still remains. It is unreasonable to expect that the society which profits actually or potentially should not share in the responsibility for what was done."[33]

A leading legal commentator, Professor Clark Havighurst, has argued similarly that requiring research agencies to bear the financial risk of adverse or unexpected effects would be in the public interest:

> The principle of societal responsibility makes not only humanitarian but economic sense, for the research industry will undertake fewer projects that are not justified by a balancing of risks (and

other costs) against potential benefits if all of the potential costs are taken into account...Whatever compensation system is devised must not only compensate the unlucky subject but also place the burden on those best able to evaluate and control the risks attending the experiment.[34]

There is now no legal rule that researchers must provide compensation for harm or indemnification for injury suffered as a result of participation in an experiment. On the other hand, researchers *are* required to state in the consent form whether or not they will provide such indemnification or compensation.[35] *Unless at least medical care for injury is provided by the researcher or hospital, patients should not consent to participate in the research.* Researchers should also provide compensation for injury, including loss of earnings. Lack of compensation is unfair to individual patients, and by participating in such research, patients help perpetuate the system that is unfair to all research subjects. One possible exception to this rule is if the patient is convinced, after full disclosure and consultation with experts not involved in the experiment, that there is no effective standard treatment available and the risks of the experiment are outweighed by its potential benefits.[36] Much research on AIDS drugs may fall into this category.[37]

What things should a patient–subject know before consenting to be a research subject in a medical experiment?

Anyone asked to be a research subject should carefully review the following questions and make sure they can answer all of them before agreeing to participate:

1. What will the experiment involve in terms of procedures done to me, inconvenience to me, and how much of my time will the experiment involve? Do I understand the procedures and their purposes?
2. What are the risks to my life, health, or bodily functions for participating in this research?
3. What benefits can I reasonably expect from participating in this research?

4. What is the purpose of this research? What benefits might others obtain if this research is successful?

5. Has this research been approved by an IRB? (If so, how many nonmedical people are on the IRB? *If not, do not participate!*)

6. Who is in charge of this research? (Make sure you talk directly with this person if you want to or need to, so the research can be properly explained to you.)

7. Is the person in charge being paid by a drug company or other for-profit company to do this research?

8. Is this a random trial (that is, will I be assigned to the group getting the experimental treatment or to the group getting a placebo, sugar pill, or standard treatment, based on random assignment like the flip of a coin)? (If so, and if you actually want the experimental treatment, you need to find out if you can get it without the random assignment that gives you only a 50 percent chance to get it.)

9. Can I have a copy of the consent form? (This is routine, but you should make sure not only that you get a copy, but also that you have time to carefully review it with a close friend. If this is a problem, do not consent!)

10. Will I be compensated for any injuries suffered as a result of participating in this experiment? (If you will not be provided with at least free medical care for injuries sustained as a result of participation, you should *not* participate.)

11. Who will have access to the results, and will the medical records with results be identifiable as my records? (If you are not satisfied that your confidentiality will be maintained, do not participate.)

12. Whom do I talk to if I have further questions? (Get a name, the qualifications and title of the person, and phone number.)

NOTES

1.   J. Fletcher, "The Evolution of the Ethics of Informed Consent," in Berg & Tranoy, eds., *Research Ethics* (New York: Alan R. Liss, 1983), at 211.

2. *Trials of War Criminals Before the Nuremberg Military Tribunals: Under Control Council Law No. 10,* vols. 1 & 2 (Washington, DC: Government Printing Office, 1949).

3. G. J. Annas, L. H. Glantz & B. F. Katz, *Informed Consent to Human Experimentation* (Cambridge, Mass: Ballinger, 1977), at 6–9.

4. *US v. Stanley,* 107 S. Ct. 3054, 3063 (1987). *See also* Shenon, "CIA Near Settlement of Lawsuits by Subjects of Mind-Control Tests," *New York Times,* Oct. 6, 1988, at A14. *And see generally* Bassiouni, Baffes & Evrard, *An Appraisal of Human Experimentation in International Law and Practice: The Need for International Regulation of Human Experimentation,* 72 J. of Crim. L. & Criminology 1597 (1981); and R. J. Lifton, *The Nazi Doctors: Medical Killing and the Psychology of Genocide* (New York: Basic Books, 1986).

5. *See generally* J. Katz, *Experimentation with Human Beings* (New York: Russell Sage Foundation, 1972); *Ethical Aspects of Experimentation with Human Subjects,* 98 Daedalus 219 (1969). Fraud and misconduct also are recurring problems ("Pressures on Medical Researchers Create Climate Conducive to Fraud," Wall Street Journal, Feb. 14, 1989 at B4).

6. Katz, *supra* note 5, at 10–65.

7. *See* Brandt, *Racism, Research and the Tuskegee Syphilis Study,* 8 Hastings Center Report 21 (Dec. 1978); and J. Jones, *Bad Blood* (New York: Basic Books, 1976). The survivors and their heirs eventually settled a lawsuit against the federal government for $37,500 each minus attorney fees. New York Times, July 29, 1979, at 26. *See also* S. Coney, *The Unfortunate Experiment* (Auckland, N.Z.: Penguin, 1988) (recounts a similar experiment in New Zealand regarding the natural history of cervical cancer).

8. Insight Team of the *Sunday Times* of London, *Suffer the Children: The Story of Thalidomide* (New York: Viking Press, 1979), at 109.

9. Beecher, *Ethics and Clinical Research,* 274 New Eng. J. Med. 1354 (1966). Beecher originally submitted 50 unethical studies, but the editors cut it to 22 "for reasons of space."

10. *E.g.,* Bernard, *Maximum Permissible Amounts of Natural Uranium in the Body, Air and Drinking Water Based on Human Experimental Data,* 1 Health Physics 288 (1958).

11. *See* Annas, *Baby Fae: The "Anything Goes" School of Human Experimentation,* 15 Hastings Center Report 15 (Feb. 1985); Jonasson & Hardy, *The Case of Baby Fae,* 254 JAMA 3358 (1985).

12. Annas, *Death and the Magic Machine: Informed Consent to the Artificial Heart,* 9 W. New Eng. L. Rev. 89 (1987).

13. *Id.* and sources cited therein. *And see* Schroeder Family, *The Bill Schroeder Story* (New York: William Morrow, 1987), at 134; and Brauer, "The Promise That Failed," New York Times Magazine, Aug. 28, 1988, at 46.

14. Prepared Statement by Professor Bernard Barber for House Subcommittee on Health, Protection of Human Subjects Act, Sept. 28, 1973; at 2; *and see*

B. Barber, J. Lally, J. Makarushka & D. Sullivan, *Research on Human Subjects* (New York: Russell Sage Foundation, 1973); and B. Barber, *Informed Consent in Medical Therapy and Research* (New Brunswick, NJ: Rutgers University Press, 1980).

15. B. Gray, *Human Subjects in Medical Experimentation* (New York: John Wiley & Sons, 1975).

16. Ingelfinger, *Informed (But Uneducated) Consent*, 287 New Eng. J. Med. 465, 466 (1972).

17. *See generally* R. Levine, *Ethics and Regulation of Clinical Research*, 2d ed. (Baltimore, Md.: Urban & Schwarzenberg, 1986).

18. 45 C.F.R. sec. 46.111(1983 rev.). For a complete review of IRB legal issues, *see* Robertson, *The Law of Institutional Review Boards*, 26 UCLA L. Rev. 484 (1979).

19. 45 C.F.R. sec. 46.107. *And see* Veatch, *Human Experimentation Committees: Professional or Representative?* 5 Hastings Center Report 31 (Oct. 1975).

20. *See* Annas et al., *supra* note 3.

21. *See* P. Ramsey, *The Patient as Person* (New Haven, Conn.: Yale U. Press, 1970), at 5.

22. *See,* e.g., Mead, *Research with Human Beings*, 98 Daedalus 361, 374 (1969).

23. R. Tittmuss, *The Gift Relationship* (New York: Pantheon Books, 1971).

24. Jonas, *Philosophical Reflections on Experimenting with Human Subjects*, 98 Daedalus 219, 245 (1969).

25. J. Katz & A. Capron, *Catastrophic Disease: Who Decides What?* (New York: Russell Sage Foundation, 1975), at 88.

26. *Id.*

27. 45 C.F.R. sec. 46.116 (1981).

28. *Id.*

29. 45 C.F.R. sec. 46.110; preliminary list published 46 Fed. Reg. 8392 (Jan. 26, 1981).

30. Annas & Glantz, *Rules for Nursing Home Research*, 315 New Eng. J. Med. 1157 (1986).

31. *Id.* and sources cited therein.

32. *Medical Malpractice: Report of the Secretary's Commission on Medical Malpractice* (DHEW Pub. No. OS 73–88, 1973), at 79.

33. Beecher, *Human Studies*, 164 Science 1256, 1257 (1969).

34. Havighurst, *Compensating Persons Injured in Human Experimentation*, 169 Science 153, 154 (1970). *See also* Calabresi, *Reflections on Medical Experimentation in Humans*, 98 Daedalus 387, 398 (1969); and Ladimer, ed., *Clinical Research Insurance*, 16 J. Chron. Dis. 1229, 1233 (1963); President's Commission for the Study of Ethical Problems in Medicine and Biomedical and Behavioral Research, *Compensating for Research Injuries* (Washington, DC: Government Printing Office, 1982); and Danzon,

"Liability and Insurance for Medical Maloccurrences: Are Innovations Different?" in M. Siegler, ed., *Medical Innovations and Bad Outcomes* (Ann Arbor, Mich.: Health Administration Press, 1987).

35. 45 C.F.R. 46.116(a)(6) (1981).

36. Extraordinarily difficult issues are presented by experimental cancer treatments and experimental AIDS treatments. Experimental cancer treatments are often urged on patients by researchers who sometimes seem more like salespeople than scientists. *See, e.g.,* Kleiman, "In Search for Cancer Cure, Patients Come Second," *New York Times,* Feb. 8, 1986, Bl; and Antman, Schnipper & Frei, *The Crisis in Clinical Cancer Research: Third-Party Insurance and Investigational Therapy,* 319 New Eng. J. Med. 46 (1988).

37. In the case of AIDS, desperate patients often "demand" experimental drugs because the only alternative to aggressive experimentation seems to be death. *See* Annas, *AIDS, Judges, and the Right to Medical Care,* 18 Hastings Center Report 20 (Aug. 1988); Levine, *Has AIDS Changed the Ethics of Human Subjects Research?,* 16 Law, Medicine & Health Care 167 (1988); Mariner, *Why Clinical Trials of AIDS Vaccines Are Premature,* 79 Am. J. Public Health 86 (1989); and Annas, *Faith (Healing), Hope, and Charity at the FDA: The Politics of AIDS Drug Trials,* Vill. L. Rev. (1989) (arguing that only scientifically valid randomized clinical drug trials are likely to demonstrate the efficacy of an ultimate treatment for AIDS and that it is harmful to AIDS patients in both the short and long term to undermine these studies by making unproven drugs available to those who can pay for them). *And see generally* Kessler, *The Regulation of Investigational Drugs,* 320 New Eng. J. Med. 281 (1989).

X

Medical Records

There are about a billion visits annually in the United States to doctors' offices, clinics, HMOs, and hospitals, and each visit either generates a new record or adds to an existing one. At the turn of the century, approximately 90 percent of all medical services were directly delivered by physicians. Today, fewer than 10 percent of all health care providers are physicians.[1] In the typical hospital, for example, one-third or less of a patient's record will be created by the attending physician.[2]

The variety of medical records is kaleidoscopic. Nevertheless, there is little dispute concerning the primary purposes for which they are kept. In the private physician's office, they are generally maintained to document the patient's history, condition, and treatment; to aid in continuity of care; and to provide a record for billing. In the hospital, Joint Commission on Accreditation of Healthcare Organizations (JCAH) standards interpret the purposes of the medical record as:

1. A basis for planning patient care and for continuity in the evaluation of the patient's condition and treatment
2. Documentary evidence of the course of the patient's medical evaluation, treatment, and change in condition
3. Documentary evidence of communication between the responsible practitioner and any other health professional contributing to the patient's care
4. Protection of the legal interests of the patient, hospital, and practitioner
5. A database for use in continuing education and research.[3]

All of these rationales can be applied to some extent to records kept in the private office, clinic, and HMO as well. Nevertheless, there are physicians who keep almost no records, and it remains fair to conclude that medical records "often fail the purposes of lucid communication, education, and rapid retrieval of stored information."[4] The growth of government financing and monitoring of medical care means that the purposes of the medical record will expand.

This chapter discusses the major legal issues that surround the making, maintenance, storage, and disposal of medical records, as well as the issue of patient access. Issues of "privacy" and third-party access to medical records are addressed in the next chapter.

What is included in a patient's record?

A physician's office records are required to conform to "accepted medical practice." Accepted practice is to maintain a record that documents the patient's history, physical findings, treatment, and course of disease.[5] Sufficient information should be included in the record to document the diagnosis and course of treatment. If a patient accuses a physician or nurse of malpractice or incompetence, the medical record will usually be the provider's best defense. This is because it was made contemporaneously with the treatment and is generally much more reliable than the memory of either the patient or the provider. It is important for both patient and provider that complete and accurate records be maintained.

Requirements regarding the maintenance and content of health care facility records are usually found in state statutes or regulations that govern licensure. These requirements fall into three groups: (1) those that simply mandate the maintenance of records that are accurate, complete, or adequate; (2) those that set forth broad categories of information that must be included; and (3) those that provide specific requirements for information that must be included. Provisions for the signing and retention of records are often contained in such regulations as well.

Although it is unlikely that a health care facility would actually lose its license for failure to comply with record-keeping requirements, licensing bodies do scrutinize the facility's compliance procedures. Moreover, hospitals can and do impose sanctions such as temporary suspension of the operating or admitting privileges on those physicians who fail to meet their record-keeping requirements.[6]

Additional requirements concerning the contents of hospital records have been promulgated by the JCAH. They include:

1. Identification information
2. Patient's medical history
3. Report of patient's physical examination
4. Diagnostic and therapeutic orders
5. Evidence of appropriate informed consent or indication of the reason for its absence
6. Observations of patient condition, including results of therapy
7. Reports of all procedures and tests, and their results. Conclusions at the end of the hospital stay or after evaluation or treatment.[7]

Is there anything that should not be included in a patient's medical record?

The debate on this question is ongoing and unresolved. On one side are those who argue that literally everything about a patient's condition and background can be relevant to proper diagnosis and treatment. On the other side are those who believe that far too much sensitive and nonessential information finds its way into medical records, to the potential detriment of the patient.

Without attempting to resolve this dispute, it can be noted that, though not illegal, it is generally considered inappropriate to include personal criticisms (for example, this patient is "fat and sloppy"; "shabbily dressed again today") or offhand comments in a patient's record ("I love her perfume"). Such statements unfairly color the attitude of the patient's next medical caretaker who reads the record. In addition, they can lead to concealment of the record from the patient, not for any legitimate reason, but simply for fear of embarrassing the person who wrote such remarks. Another reason casual comments should not be included is that the medical record may be viewed by many other people during its lifetime. These casual comments may be used against the patient by potential schools, employers, insurers, or governmental agencies.

Providers are counseled to record *facts* about a patient (for example, "speech slurred, eyes bloodshot") rather than conclusions from these observations that may not be true (for example, "patient is an alcoholic").

Who owns the patient's medical records?

The *information* contained in the record can be properly characterized as the patient's property. Nonetheless, the general rule is that the owner of the paper on which the medical record is written is the "owner" of the record. Some states even have statutes that specify that health care facilities "own" the medical records in their custody. Likewise, physicians, even if not covered by statute, can properly be considered the owners of the medical records generated and maintained by them in their private offices.

What rights does the owner of a medical record have?

Ownership of a medical record is a limited right that is primarily custodial in nature. Possession of the record is governed by many other statutes (such as licensing statutes) and contracts (such as health insurance contracts) as well as by the interests of the patient in the contents of the medical record. Providers have custody of records and strong interests in them, but the patient has an even stronger interest. The patient's interest is strong enough to give the patient a legal and ethical right of access to the information contained in the records and a right to a complete copy of the records themselves.

Why might a patient want to read the medical record?

The primary reason a patient might want to read the medical record is so that the patient can better understand his or her medical condition and cooperate in treatment. Other reasons include: to check the accuracy of family and personal histories, to be better informed when asked to consent to diagnostic and therapeutic procedures, to better understand the role the physician is taking in the treatment, and to appreciate more fully the state of one's health to be better able to prevent a recurrence of the disease or condition in the future. The medical record can be a powerful means of health education that can benefit the patient while in the hospital and after discharge.

If a patient is moving out of town, out of the country, or going on a long trip, the patient may also have a good reason to take along a copy of the medical record or at least the discharge summary of the patient's most recent hospital visit.

There are many parts of the record (which could cover hundreds of pages), such as laboratory reports, in which the patient may have little or no interest. When having copies made, therefore, the patient should try to review the entire record first and order copies only of those pages the patient needs to answer specific questions.

Will the patient be able to understand the medical record without the help of a physician?

If the patient can decipher the handwriting in the record, knows the meaning of the abbreviations used, and has a good medical dictionary, the patient can get some important information out of the record.

As for reading the handwriting, even a physician may not be able to help. It is common practice among physicians and epidemiologists who do research with medical records to hire someone to interpret and type them before the research study is conducted. The meaning of the abbreviations used varies somewhat from hospital to hospital, and a staff member from the particular hospital can explain what the various abbreviations mean. Even without such help, patients should be able to check the accuracy of the history they related, the diagnosis, test results, and treatments prescribed. Detailed record analysis, however, requires consultation with an experienced medical care provider or nurse.

Does a patient have the legal right to see and copy the medical record?

The majority of states grant individuals a *legal* right to see and copy their medical records by statute, regulation, or judicial decision.[8] In most other states, individuals have a probable legal right to access without bringing suit. In *all* states, patients *should* routinely be provided access regardless of the law because the information in the medical record is about them, they have the strongest personal interest in the information, and they may need the information to decide on treatment, whether to change physicians, and how to plan for their future.

In some states, the medical records access statute applies only after discharge. Some state statutes also have limitations on access to psychiatric records. Others limit the types of records that are available. Some statutes, for example, exclude access to laboratory reports, X-rays, prescriptions, and other technical information used in assessing the patient's health condition. A few state statutes require the patient to show "good cause" before the patient has a right to read the record, but what is meant by this term is not specified in the statute. Other states have more limited statutes that either provide for access under specified circumstances or require the patient to obtain access through an attorney, physician, or relative.

State medical licensing boards are also beginning to recognize the importance of patients having access to their medical records.[9] On the federal level (for example, Department of Veterans' Affairs facilities), the Privacy Act of 1974 requires direct access in most circumstances. The Privacy Protection Study Commission, established by that act, recommended, "Upon request, an individual who is the subject of a medical record maintained by a medical care provider, or another responsible person designated by the individual, be allowed access to that medical record including an opportunity to see and copy it." [10]

This recommendation is significantly broader than one made more than fifteen years ago by HHS's Medical Malpractice Commission. It recommended simply that "states enact legislation enabling patients to obtain access to the information contained in their medical records through their legal representatives, public or private, without having to file a suit." [11]

Why is it that some health care facilities and doctors do not want patients to see medical records?

Information is power, and some doctors simply do not want to share decision-making power with patients. Others view patients like children and complain that patients will misunderstand the information or get upset by it. A classic example of physician distaste for patients who insist on reading their medical records is a study on "record reading" by four psychiatrists at a major Boston teaching hospital. The psychiatrists interviewed the 11 out of 2500 consecutive patients at the hospital who, in a one-year period, asked to see their medical records.[12]

The authors concluded that patients who ask to see their medical records have a variety of personality defects, usually manifesting themselves in mistrust of, and hostility toward, the hospital staff. But to locate the source of mistrust in the patient's personality, rather than in the stress of illness and hospitalization, is to forget, as psychiatrist Donald Lipsett perceptively notes, that "the doctor–patient relationship cannot be understood simply in terms of the patient's side of the equation."[13] The authors of this study of patients who ask to see their medical records thus fell into what Robert Burt of Yale Law School has referred to as "the conceptual trap of attempting to transform two-party relationships, in which mutual self-delineations are inherently confused and intertwined, by conceptually obliterating one party."[14] Thus, it would seem that the ten women at the Boston hospital who asked to read their charts "to confirm the belief that the staff harbored negative personal attitudes toward them" were correct in their belief; the psychiatrists had in fact labeled them "of the hysterical type with demanding, histrionic behavior and emotional over-involvement with the staff."

The authors of the Boston study also seem to have been unaware of the wide variety of settings in which patients have *benefited* from routine record access (and incorrectly asserted that there were no strikingly beneficial effects in the two studies they did cite). In one of these previous studies, for example, two patients expressed their unfounded fear that they had cancer only after their records were reviewed with them, and one pregnant patient noted an incorrect Rh typing that permitted Rho-Gam to be administered at the time of delivery.[15] In the other study, half of the patients made at least one factual correction in their records.[16]

In short, the Boston study seems to have been done primarily to try to prove that the right of access to medical records is unimportant, since it is only exercised by "mentally disturbed" people who are not improved by reading their charts. The study failed to prove this. Even if it had succeeded, we should still be unwilling to deprive the other 2489 patients at that hospital of their right to have access to their medical records in the future. If we believe in individual freedom and self-determination, we must give all citizens the right to make their *own* decisions and to have access to the same information that is widely available to those making decisions about them. Shared decision making is impossible without shared information. It is, moreover, as irrelevant in this connection that 2489 patients at one hospital did not ask to see their records as it is that more than 200 million Americans never have had to exercise their right to remain silent when arrested. Rights serve us all, whether we exercise them or not.

What are the primary problems patients have when they read their medical records?

The primary problems patients have with medical records involve interpreting medical abbreviations, vocabulary, and physicians' handwriting. In places where access has been available for years, no adverse physical or mental reactions have been reported.

In Massachusetts, for example, patient access has been mandated by statute since 1946 without a single reported adverse incident. And the Privacy Study Commission, hearing from all federal agencies that had adopted access policies under the Privacy Act of 1974, found that "not one witness was able to identify an instance where access to records has had an untoward effect on a patient's medical condition." [17] Finally, a two-year study at the outpatient department of Boston's Beth Israel Hospital of routine record sharing with patients concluded that physicians' fears about liberal patient access to records were unwarranted, and that the open access policy made the relationship between patients and professionals "more collaborative." [18]

Some physicians have suggested giving all patients carbon copies of their records as soon as the records are made. The advantages of this system include:

1. Increased patient information and education
2. Continuity of records as patients move or change physicians
3. An added criterion on which patients may base selection of physicians
4. Improvement in the doctor–patient relationship by making it more open
5. An added way for physicians to monitor quality of care
6. Increased responsiveness to consumer needs [19]

This proposal deserves serious consideration. The burden of proof is now squarely on those who would deny patients access to their records to demonstrate other than paternalistic and self-serving reasons for this policy.

Might disclosure of medical records be harmful to the physician?

This possibility is considered likely by some medical commentators, but a general policy of open access creates far fewer potential legal problems than a general policy of denying patients access to their records. Physician–commentator Michael J. Halberstam has unequivocally advised against patient access, arguing "the chart is none of the patient's business." He further advised concealing mistakes from patients (at least those that do not seem to have caused any serious damage) and suggested that the medical record is "a time bomb lying in wait to give physicians trouble two or three years in the future." [20]

Such sentiments tend to perpetuate the public's view that health care providers are trying to hide something from them. Policies based on these sentiments lead to cynicism about the potential for peer review and increase the likelihood that a lawsuit will be filed primarily for the purposes of gaining access to medical records.

There is also evidence that a policy of open records changes the content of the records themselves. Such changes are usually improvements and include the deletion of personal comments, more detailed documentation of informed consent, and better handwriting. Patients should encourage their health care providers to work with them in a partnership characterized by openness and trust, not paranoia and secrecy.

When can the physician or health care facility refuse to allow a patient access to the medical record?

If the patient has a statutory right to the record, or if the patient brings suit, there is nothing that can legally prevent the patient's access. Inherent in the law, however, is the concept of reasonableness. It may be reasonable for a health care facility to suggest that the patient's physician be present while the patient inspects the record. On the other hand, it is unreasonable to require this in most circumstances. For instance, it is unreasonable to deny the patient access to records if the patient must make a treatment decision and needs the information to do so. A facility or physician may not make it impossible for the patient to exercise the right by imposing an exorbitant charge for copying.

Some of the statutes provide, and common sense dictates, that if the physician has good reason to believe that access to the record will be harmful to the patient, direct access by the patient may be denied. But this must not be treated casually. The physician must be reasonable in this belief and should be able to document this belief on the basis of objective evidence. The patient should also be given the opportunity to challenge this decision and to designate another person to receive the information on the patient's behalf.[21]

How can a patient have mistakes in the medical record corrected?

The patient should speak to the person who wrote the incorrect entry and ask that it be corrected. One of the legal lessons most health professionals have learned is *never alter a medical record*. This is excellent advice, since a jury will often consider an altered medical record as the equivalent of admitting negligence. Nevertheless, like almost every rule, this one has some obvious exceptions: The most obvious is when the record is incorrect. There have been many suggestions concerning record alteration. Perhaps the most widely recommended is to cross out the incorrect information in such a way that it is still legible, write in the correct information, and add a dated note that explains why the information was changed (for example, "changed because it was later discovered to be inaccurate").[22] Such a procedure maintains the integrity of the medical record and permits the correction of false data. Each physician and health facility should have a procedure for making record corrections. In the event that a patient asks that the record be corrected, and the health care provider determines that no correction is warranted, the patient should at least be permitted to have his or her version of the facts added to the record.

How long will a doctor, health care facility, or HMO retain a copy of the patient's medical record?

This will vary from facility to facility and doctor to doctor, although the trend is to retain medical records forever. More than two-thirds of the states have statutes or regulations that provide a specific minimum length of time for the retention of hospital records. These vary from two years to thirty years.[23]

The following recommendations have been made to health care providers regarding retention of medical records:

1. In the interests of medical science and good patient care, medical records should be retained for as long as possible.
2. At a minimum, records should be retained:
   a. if the patient is a competent adult—for the longest statute of limitations that may apply, usually six years.
   b. if the patient is an incompetent adult, or becomes incompetent before six years have expired—until the patient recovers plus the remaining statutory time, or another time prescribed by statute for this specific circumstance.
   c. if the patient is a minor—until the patient reaches the age of majority plus the statutory period for malpractice actions, or another period specified by statute for this circumstance.
3. If the physician finds that her facilities preclude the retention of records for any longer than the recommended minimum periods, it is recommended that she notify the patient that his records will no longer be retained, and give the patient or the patient's designee the copy of the record upon payment of a reasonable fee for postage and handling.
4. Microfilm reproductions are as fully admissible in most courts as the original. However, it is advisable to retain the original records for the recommended minimum periods, since the original is more convenient to read and handle and presence of the original minimizes the possibility of assertions that it has been altered or is incomplete.[24]

Retirement or death of the physician does not alter the confidential nature of medical records, nor does it relieve the physician or the estate of any professional liability for which the physician would otherwise be accountable. Thus, the physician or the estate must continue to safeguard the confidentiality of medical records.

The records, as long as they exist, are confidential, and the patient continues to have an interest in the information contained therein. They cannot, for example, be sold without the patient's permission.[25]

What should the patient do if denied access to the medical record?

Raise hell. There is no valid ethical or legal reason to deny a competent patient access to the medical record. If in the hospital, the patient can (and should) refuse to consent to further treatment or testing until permitted to review his or her medical record. In addition, the patient should lodge complaints with the hospital's patient representative, hospital administrator, and the institution's ethics committee. It is almost certain that this will result in access, since denial is an old fashioned idea that few US physicians and administrators consider reasonable medical practice.[26]

In extreme cases, the attending physician may believe that access will be harmful to the patient's health and may not want to take personal responsibility for this potential harm. Although this explanation seems bizarre, the patient should be prepared to compromise by permitting the physician to give a complete copy of the record to the patient's advocate or to another individual designated by the patient (a friend or relative). That person can then take the "responsibility" for providing the copy directly to the patient. Parents should have routine access to the records of their children if they are consenting to their medical treatment, and guardians should have routine access to the medical records of their wards.

Access may be more difficult to obtain after discharge, however, for the simple reason that it is so easy for the medical records librarian to stonewall requests. Nonetheless, going directly to the office of the hospital administrator and demanding a copy of the record is usually effective. If this does not work, the help of the state attorney general, Office of Consumer Affairs, or Department of Health should be sought. If none of these public agencies will help, then the services of an attorney may have to be enlisted to gain access to the medical records. Usually, all that will be required is the patient's signature on a form releasing the records to the patient and a fee for copying them. *Remember*: By requesting medical records, patients not only help educate themselves about their medical condition but also help every other person who requests their own medical record in the future. The more people who make this request, the more routine and ordinary it becomes, and the less likely health care providers will continue to label "record-seeking behavior" as aberrant or strange.

NOTES

1. V. Fuchs, *Who Shall Live?* (New York: Basic Books, 1974), at 56.
2. Privacy Protection Study Commission, *Personal Privacy in an Information Society* (Washington, DC: Government Printing Office, 1977), at 278.
3. Joint Commission on Accreditation of Healthcare Organizations, *Accreditation Manual for Hospitals*, 1989 ed. (Chicago: JCAH, 1988), at 95.
4. Goldfinger & Dineen, "Problem-Oriented Medical Records," in Thorn *et al.*, eds., *Harrison's Principles of Internal Medicine*, 8ᵗʰ ed. (New York: McGraw-Hill, 1977), at 9.
5. *See, e.g.,* "Approach to Disease," in *Harrison's, supra* note 4; and Dobson, *Documenting Outpatient Medical Records*, 75 Federation Bull. 323 (1988).
6. *See, e.g., Board of Trustees of Mem. Hosp. v. Pratt*, 72 Wyo. 120, 262 P.2d 682 (1953), upholding the right of a hospital to suspend the privileges of a physician who did not comply with the requirement to keep records up-to-date. And *see* Grayson, *Incomplete Medical Records: Three Case Solutions*, 8 Hospital Medical Staff (March 1979).
7. *Accreditation Manual for Hospitals, supra* note 3, at 97.
8. For a state-by-state listing of the current statutes and a summary of the judicial decisions on medical records, *see* "Medical Records," in Health Law Center, *Hospital Law Manual* (Rockville, Md.: Aspen, 1988), at 78–107 (updated periodically); *see also* W. H. Roach, S. N. Chernoff & C. L. Esley, *Medical Records and the Law* (Rockville, Md.: Aspen, 1985), at 219–72; Kaiser, *Patient's Rights of Access to their own Medical Records: The Need for New Law*, 24 Buffalo L. Rev. 317 (1974); and Note, *Toward a Uniform Right to Medical Records: A Proposal for a Model Patient Access and Information Practices Statute*, 30 UCLA L. Rev. 1349 (1983). Information on state law can be obtained from the state attorney general's office, or by writing the American Medical Records Association, 875 N. Michigan Ave., Chicago, Ill. 60611
9. *E.g.,* the Massachusetts regulation (243 CMR 2.06 [13]), reads in part: A licensee shall provide a patient or, upon a patient's request, another licensee or another specifically authorized person, with the following:

i. A summary, which includes all relevant data of that portion of the patient's medical record which is in the licensee's possession, or a copy of that portion of the patient's entire medical record which is in the licensee's possession. It is within the licensee's discretion to determine whether to make available a summary or a copy of the entire medical record.

ii. A copy of any previously completed report required for third party reimbursement.

10. *Personal Privacy, supra* note 2, at 298.

11. *Secretary's Report on Medical Malpractice* (HEW, DHEW Pub. No.OS 73-88, 1973), at 77.

12. Altman, Reich, Kelly & Rogers, *Patients Read Their Hospital Charts,* 302 New Eng. J. Med. 169 (1980).

13. Lipsitt, *The Patient and the Record,* 302 New Eng. J. Med. 169 (1980). And *see* letters to the editor, 302 New Eng. J. Med. 1483-84(1980); Roth, Woldord & Meisel, *Patient Access to Records*: Tonic or Toxin? 137 Am. J. Psychiatry 592 (May 1980); and *Pierce v. Penman,* 515 A.2d 948 (Pa. Super. Ct. 1986).

14. R. Burt, *Taking Care of Strangers* (New York: Free Press, 1979), at 113.

15. Golodetz, Ruess & Milhous, *The Right to Know: Giving the Patient His Medical Record,* 57 Arch. Phys. Med. Rehab. 78 (1976).

16. Stevens, Stagg & MacKay, *What Happens When Hospitalized Patients See Their Own Records,* 86 Annals Internal Med. 474 (1977) *And see* Stein *et al., Patient Access to Medical Records on a Psychiat ric Inpatient Unit,* 136 Am. J. Psychiatry 3 (1979); Bouchard *et al.,* reported in "How to Reduce Patient's Anxiety: Show Them Their Hospital Records," *Medical World News,* Jan. 13, 1975, at 48; and Gill & Scott, *Can Patients Benefit from Reading Copies of Their Doctor's Letters about Them?,* 293 British Med. J. 1278 (1986).

17. *Personal Privacy, supra* note 2, at 297.

18. Knox, "Medical Board Told Patients Should Get Access to Records,"Boston Globe, Mar. 2, 1978, at 15.

19. Shenkin & Warner, *Giving the Patient His Medical Record: A Proposal to Improve the System,* 289 New Eng. J. Med. 688 (1973). *See also* letters to the editor, 290 New Eng. J. Med. 287–88 (1974)

20. Halberstam, "The Patient's Chart Is None of the Patient's Business," *Modern Medicine* Nov. 1, 1976, at 85–86

21. *E.g.,* 243 CMR 2.06 (13) (d) (Massachusetts Board of Medicine Regulation). More recently it has been suggested that hospitals are making it increasingly difficult for patients and their physicians to obtain copies of records and radiographs. To help make access easier, and diminish the need for unnecessary repetition of invasive studies, deposit charges for radiographs should be eliminated, copying fees should be reduced, and requests for unabridged medical records should be honored and processed in no

more than forty-eight hours from receipt "if the records [as is increasingly the case with commercial medical-record depositories] are not stored on the hospital grounds" (Kaufman, "Barriers Between Patients and Their Medical Records" [letter], 319 New Eng. J. Med. 1672 [1988]).

22. *See, e.g., Hospital Law Mannual*, supra note 8, at 16.

23. *Id.*, and *see* supra note 8.

24. G. J. Annas, L. H. Glantz & B. F. Katz, *The Rights of Doctors, Nurses and Allied Health Professionals* (Cambridge, Mass: Ballinger 1981), at 163–64.

25. *Id.*

26. It may take the rest of the world longer to adopt an open medical records policy. For example, it was still seen as radical for patients in New Zealand to ask to see their medical records in 1987. S. Coney, *The Unfortunate Experiment* (Auckland, N. Z.: Penguin, 1988), at 8. In the United States, on the other hand, there are even commercial companies to assist individuals in organizing and maintaining copies of their medical records, such as Medic Alert (1-800-ID-ALERT) and the American Academy of Pediatrics' Medical Card (312-228-5005). Individuals are increasingly being advised to travel with a copy of the results of their latest physical examination or hospital discharge summary, and individuals with heart conditions should carry a copy of their cardiogram when they are away from home. *See* Sloane, "Keeping a Medical History in Hand," *New York Times*, Jan. 14, 1989, at 50.

XI
Privacy and Confidentiality

As society becomes more and more dependent on a galaxy of information systems, two conflicting trends emerge. The first trend, exemplified by state and federal Freedom of Information or Sunshine acts, is to provide the public access to all information held by governmental agencies. The premise is that public knowledge of the most intimate details of how government works is likely to make government more responsive to the will of the people and to prevent official wrongdoing (such as trading arms for hostages). The second trend is exemplified by state and federal laws (such as the federal Privacy Act) designed to protect information about individual citizens from public disclosure. Details remain to be worked out in many areas, but the consensus is that with all forms of personal data-keeping systems, such as credit, insurance, education, taxation, criminal, and medical, individuals have or should have a right to examine and correct the information and, under most circumstances, to prevent its release without their knowledge and express consent.

Medical records have been the last to come under public scrutiny, perhaps because medicine has a tradition of "keeping confidences." But now that sole practitioners have become an endangered species, record keeping in medicine resembles other massive record-keeping systems. Accordingly, the rules applied to these other systems will likely be applied to medical records as well. The concept of confidentiality of medical records has been much more discussed than litigated, and only a few dozen cases have reached the appellate court level. The law in this area is still in its infancy, and resort must often be made to public policy arguments and analogy.

This chapter continues the discussion of medical records begun in the previous chapter, focusing on the legal concepts of privacy and confidentiality, the exceptions to these doctrines, and the considerations that should go into a decision about releasing confidential medical information to anyone other than the patient.

175

What is meant by the terms "confidentiality," "privilege," and "privacy?"

Almost all law dealing with access to medical records by persons other than the patient can be categorized under the headings of confidentiality, privilege, and privacy. As commonly used, to tell something in confidence means that the person will not repeat the information to anyone else. *Confidentiality* presupposes that something "secret" will be told to a second party (such as a doctor) who will not repeat it to a third party (such as an employer). Relationships such as attorney–client, priest–penitent, and doctor–patient are confidential relationships. In the doctor–patient context, confidentiality is understood as an expressed or implied agreement that the doctor will not disclose information received from the patient to anyone not directly involved in the patient's care and treatment.

A communication is privileged if the person to whom the information is given is forbidden by law from disclosing it in a court proceeding without the consent of the person who provided it. *Privilege*, sometimes called "testimonial privilege," is a legal rule of evidence, applying only in the judicial context. The privilege belongs to the client, not to the professional, although the hospital and physician may have a duty to assert it on behalf of the patient. Unlike the attorney–client privilege, the doctor–patient privilege is not recognized at common law and therefore exists only if a state statute establishes it (most states have such statutes).

There are at least two senses in which the term *privacy* is generally used. The first describes a constitutional right to privacy. Found in the liberty interests protected by the Fourteenth Amendment, this right to privacy is the basis for the opinions by the US Supreme Court limiting state interference with individual decisions about birth control and abortion. It specifically relates to an individual's ability to make important, intimate decisions that affect one's personhood free from government interference.

In the more traditional sense, the right to privacy has been defined as "the right to be let alone, to be free of prying, peeping, and snooping," and as "the right of someone to keep information about himself or his personality inaccessible to others."[1]

Is the maintenance of confidentiality a legal or an ethical obligation of health care providers?

It is both. Historically, the doctrine was an ethical duty applicable only to physicians. Currently, it is also a legal duty of physicians (as enunciated in case law, and some state-licensing statutes), and it is becoming a legal duty of other health care practitioners as well.

The Hippocratic Oath sets out the duty of confidentiality in the following words: "Whatsoever things I see or hear concerning the life of man, in any attendance on the sick or even apart therefrom, which ought not to be noised about, I will keep silent thereon, counting such things to be professional secrets."

This oath has been reinterpreted in section 4 of the American Medical Association's Principles of Ethics: "A physician shall respect the rights of patients, of colleagues, and of other health professionals, and shall safeguard patient confidences within the constraints of the law."

Similarly, section 2 of the American Nurses' Association Code provides, "The nurse safeguards the client's right to privacy by judiciously protecting information of a confidential nature." The interpretive statement elaborates:

> When knowledge gained in confidence is relevant or essential to others involved in planning or implementing the client's care, professional judgment is used in sharing it. Only information pertinent to a client's treatment and welfare is disclosed and only to those directly concerned with the client's care....The nurse–client relationship is built on trust. This relationship could be destroyed and the client's welfare and reputation jeopardized by injudicious disclosure of information provided in confidence.

The reason for all of these rules is that health care providers must often know the most personal and possibly embarrassing details of the patient's life in order to help the patient. Patients are not likely to disclose these details freely unless they are certain that no one else, not directly involved in their care, will learn of them. As one court described the patient's dilemma:

> Since the layman is unfamiliar with the road to recovery, he cannot sift the circumstances of his life and habits to determine what is

information pertinent to his health. As a consequence, he must disclose all information in his consultations with his doctor, even that which is embarrassing, disgraceful, or incriminating. To promote full disclosure, the medical profession extends the promise of secrecy. The candor which this promise elicits is necessary to the effective pursuit of health; there can be no reticence, no reservation, no reluctance when patients discuss their problems with their doctors.[2]

Is it realistic for hospital patients to think that medical information about them will remain secret or "confidential?"

No. The old "rule" in the hospital was that "everyone has access to the patient's medical record except the patient"; the modern rule is that everyone (including the patient) has access. This is the reality, although some efforts are made to restrict access. As Lawrence Altman, a physician and medical writer for the *New York Times* has correctly noted, "It has become increasingly apparent that patients can no longer assume that intimacies about their bodies or their private lives will be held in confidence by their physicians and those who work with those physicians."[3] Altman discusses a major teaching hospital (which is far from unique) that felt it necessary to post signs in its elevators saying: "Hospital staff are reminded that patient information should not be discussed in public areas."

Information exchange in the hospital is the product of the team approach to medical care in that setting, as well as use of the medical record for educational, financial, and quality monitoring purposes. When a patient of his in a general hospital surgical bed asked how many people had access to his medical record, physician Mark Siegler decided to find out.[4] It was not easy. He concluded that at least seventy-five people at the hospital had a *legitimate* need to have access. These persons included the patient's six attending physicians, twelve house officers, twenty nursing personnel (on three shifts), six respiratory therapists, three nutritionists, two clinical pharmacologists, four unit secretaries, fifteen students, four hospital financial officers, and four chart reviewers (utilization review, quality assurance review, tissue review, and the insurance auditor). Information access should be limited to a "need-to-know" basis. But even on this basis, the reality is that many people will fit into this category and thus have free access to the patient's record. Patients should obviously be told what doctors mean by "medical confidentiality." As described in the medical codes, Siegler concluded, confidentiality is a "decrepit concept," a "myth" that we should spend no more energy perpetuating.

This reality is why many patients fear having very sensitive information about them entered in the hospital chart, such as a psychiatric diagnosis or the diagnosis of AIDS or HIV infection. They know this information will spread rapidly in the hospital and could adversely affect the way in which they are treated by some members of the hospital staff. It could also "leak" from the hospital, affecting their housing, employment, and insurance status. Routine access by so many people makes it imperative that patients have access to their records so they can correct mistakes that may directly affect their care and their future.[5]

What types of incidents involving confidentiality get to court?

There are very few reported appellate cases involving breaches of confidentiality. This can mean that patients seldom learn of violations when they occur, that patients do not think it is appropriate to sue for such violations (because of the cost, uncertain damages, and possible further publicity of the confidential information), or that almost all such cases are settled before they reach an appellate court.

Those cases that have been appealed have most often alleged violation of confidences by physicians in one of the following situations: disclosure to a spouse (involving either a disease related to the marriage or a condition relevant in a divorce, alimony, or custody action); disclosure to an insurance company; or disclosure to an employer.[6]

What function does the testimonial privilege serve?

The privilege not to have to testify is founded on the belief that certain types of relationships are potentially so beneficial to individuals and society that they should be fostered by forbidding in-court disclosure of the informational content of the relationship. A privilege is therefore granted to encourage the employment of professionals by individuals who need their services and to promote absolute freedom of communication in the relationship. The contrary principle is that the courtroom is a place for the discovery of truth, and no reliable source of truth should be beyond the reach of the court. At common law, the court's interest in the truth routinely won out, and physicians were forced to testify about what was disclosed to them by their patients.

The great majority of courts currently agree that the principal reason for the privilege is to encourage a patient to freely and frankly reveal to a physician all the facts and symptoms concerning the patient's condition so that the physician will be in the best possible position to correctly diagnose and successfully treat the patient. Most states have adopted the privilege by statute. On the other hand, there are so many exceptions to the privilege rule that it rarely frustrates justice by withholding the truth from the court.[7]

Where does the "right to privacy" come from?

In *Privacy and Freedom*, Alan Westin defines privacy as "the claim of individuals, groups, or institutions to determine for themselves when, how, and to what extent information about them is communicated to others."[8] He goes on to argue that, as thus defined, the concept has its roots in the territorial behavior of animals, and its importance can be seen to some extent throughout the history of civilization. Specific protections of privacy were built into the Constitution by the framers in terms that were important to their era. With the subsequent inventions of the telephone, radio, television, and computer systems, more sophisticated legal doctrines were developed in an attempt to protect the informational privacy of the individual. Many diverse acts come under the heading of privacy violations, but most involving medical records are in the area generally described as the "publication [disclosing to one or more unauthorized persons] of private matters violating ordinary decencies."

A court can conclude that the unauthorized disclosure of medical records is an actionable invasion of privacy even without a state statute that specifically forbids it. As an Alabama court put it in a case involving disclosure of medical information by a physician to a patient's employer: "Unauthorized disclosure of intimate details of a patient's health may amount to unwarranted publication of one's private affairs with which the public has no legitimate concern, such as to cause outrage, mental suffering, shame, or humiliation to a person of ordinary sensibilities."[9]

The policy underlying the right is that because of the potential severe consequences to individuals, certain information about them (such as their HIV status) should not be repeated without their permission. In the words of one legal commentator: "The basic attribute of an effective right of privacy is the individual's ability to control the flow of information concerning or describing him."[10] Most of the cases in the doctor–patient context alleging violation of the right to privacy have involved actions in which personal medical information has been published in a newspaper or magazine, and often the suit is against the publisher rather than the physician.[11]

Under what circumstances *must* a health care provider report confidential medical information about the patient?

The doctrines of confidentiality and privacy are, on the surface, very powerful legal tools. Their effect in the day-to-day practice of medicine, however, is diluted both by the reality of the hospital and by the exceptions and defenses physicians and other health care providers can raise to a charge of unauthorized disclosure. There are even times (described in the four following circumstances) when health care professionals must disclose confidential information, even though the patient may not want them to, and the provider may not want to:

1. Public-Reporting Statutes

Almost all states have statutes that require physicians to report certain listed conditions and diseases to public authorities. These fall into four major categories: vital statistics, contagious and dangerous diseases, child neglect and abuse, and criminally inflicted injuries. These statutes decree a public policy that takes precedence over the health care professional's obligation to maintain patient confidences:

    a.   Birth and death certificates must be filed. In the case of birth certificates, information about the parents is generally required. If death is sudden or from an accidental cause, or if foul play is suspected, the medical examiner or coroner is usually required by law to do an autopsy and file a complete report with the district attorney.

b. Infectious, contagious, or communicable diseases must usually be reported. The California statute, for instance, lists cholera, plague, yellow fever, malaria, leprosy, diphtheria, scarlet fever, smallpox, typhus fever, typhoid fever, paratyphoid fever, anthrax, glanders, epidemic cerebrospinal meningitis, tuberculosis, pneumonia, dysentery, erysipelas, hookworm, trachoma, dengue, tetanus, measles, German measles, chicken pox, whooping cough, mumps, pellagra, beriberi, Rocky Mountain spotted fever, syphilis, gonorrhea, rabies, and poliomyelitis.[12] AIDS, but not HIV status, is reportable in all states. Most public health officials agree that only a fraction of many of these diseases are reported by physicians.

c. Child abuse cases must be reported. Such reporting is most common in the emergency rooms of large city hospitals. Failure to report can also subject a health professional to criminal penalties (most statutes also require nurses, social workers, teachers, and others to report).[13]

d. Specific types of injuries must be reported, such as "a bullet wound, gunshot wound, powder burn or any other injury arising from or caused by the discharge of a gun or firearm, and every case of a wound which is likely to or may result in death and is actually or apparently inflicted by a knife, icepick or other sharp instrument."[14]

2. Judicial Process

When a person makes his or her own physical condition an issue in a lawsuit (for example, a personal injury claim), most courts permit examination of the person's physician under oath either before or during the trial.[15] Even in states that have a privilege statute, there are many exceptions that permit medical information to be used in court.[16]

3. Statutes on Cost and Quality Control

There are a variety of state and federal statutes that permit certain monitoring agencies to have access to patient records for such purposes as peer review, utilization review, studies to protect against provider-reimbursement fraud, and licensing and accreditation surveys. The most pervasive of these is the federal statute and regulations on Professional Review Organizations (PROs). Regulations require that the information collected under this program not be made public in any way in which individual practitioners or patients can be identified.

4. Patient Poses a Danger to Known Person

Until the mid-1970s, physicians were held responsible for injuries inflicted on others by their patients only when the physician was directly negligent in the treatment of the patient. For example, physicians have been held responsible for negligently failing to diagnose tuberculosis, thereby placing family members at risk, and for wrongly informing a patient's neighbor that smallpox was not contagious.[17] In addition, physicians have been responsible for wrongly informing the members of a family that typhoid fever and scarlet fever of a sibling would not infect other members of the family.[18]

The California Supreme Court has gone somewhat further, ruling that "when a doctor or a psychotherapist, in the exercise of his professional skill and knowledge, determines, or *should determine*, that a warning is essential to avert danger arising from the medical or psychological condition of his patient, he incurs a *legal obligation* to give that warning" (emphasis added).[19] The case involved a patient who had threatened to kill his former girlfriend. The therapist believed him, took some initial steps to have him confined, and then, allegedly on orders from his superior, dropped the case. The patient in fact killed the young woman, and her family sued the therapist. The California Supreme Court declared, in the above-quoted language, that the therapist could be found liable if he had failed to warn the intended victim or failed to take other steps, such as confining the patient to a mental institution, to prevent the murder. The case was eventually settled out of court.

The psychiatric profession argued vehemently against this ruling. Psychiatrists predicted that the decision would curtail the ability of psychiatrists to treat patients (who would not come to them because they feared being reported to the authorities). It would also, some believed, encourage psychiatrists to have potentially dangerous patients committed to mental institutions rather than take a chance of continuing to treat them on an outpatient basis.[20] The broader we as a society make the health care professional's mandate to report the patient's condition to others, the more like a policeman the professional becomes. Nevertheless, the California Supreme Court's decision that health care providers have an obligation to society as well as to their patients is correct in life-and-death situations and has been followed by other courts.[21] When the life of another can be saved by breaching a confidence, and there is no reasonable alternative for accomplishing the same objective, courts (and society in general) will properly have little difficulty mandating disclosure.

In the absence of a statute that forbids disclosure, this interest in protecting innocent potential victims of one's patient would require a physician to contact known sexual partners of patients infected with HIV upon diagnosis, unless the patient credibly agreed to disclose this information or was able to persuade the physician that there was no risk to the sexual partner (because they no longer were having sex or were practicing "safe sex").[22]

When can health care providers, at their own option, disclose confidential information about a patient?

"Optional" disclosure situations are extremely broad. The following four situations indicate the vast discretion courts are likely to afford health care professionals who act in good faith:

1. Implied Consent
This is probably the major cause of leakage from the medical records system. The patient in a health care facility implicitly consents to the viewing of medical record by all those directly concerned with the patient's care. As noted previously, this may include the nurses on all three shifts, the ward secretary, all medical students, interns, and residents, the attending physicians and consultants, and

perhaps social, psychological, medical, or psychiatric researchers.[23] All of this may take place without the patient being made aware of it, and is a matter of custom.

2. General Release Forms (Consent)

On entering a health care facility, the patient is asked to sign a variety of forms. One of these is likely to be an authorization that essentially states that the facility may release medical information concerning the patient to anyone it thinks should have it or to certain named agencies or organizations. This will likely include insurance companies and the welfare department (if they are paying all or part of the bill) and other agencies and individuals monitoring quality and cost. No restriction is generally placed on the amount of information that may be released or the use to which these parties may put the informtion. Receivers should, however, be liable for an invasion-of-privacy action if they use the medical information for other than the specific purpose for which the health care facility released it to them.

Most general release forms can be justifiably criticized as unduly broad and so vague that the patient cannot reasonably and know-ingly sign them. The cases invalidating vague blanket surgical con-sent forms, which give the doctor and hospital authority to perform whatever procedures they think necessary, support this line of rea-soning. Another argument is that the patient's lack of bargaining power makes the signature on the form involuntary and thus ineffec-tive (for example, a sick patient may need admission or insurance coverage and cannot afford to forgo it by refusing to sign a required form).[24]

3. Private Interests of the Patient

Courts afford physicians generous latitude in making disclosures that physicians believe in good faith are in the best interests of their patients. This rule, for example, is used to justify many disclosures to spouses and close relatives without the patient's consent. When only the patient's individual welfare is involved, the patient should have the exclusive right to decide if and when confidential informa-tion should be released. Nevertheless, courts will probably continue to give physicians and hospitals much discretion in this area, so long as they can make a reasonable argument that there was no reasonable alternative action, and that the patient's health required the disclosure.

Even though there has been little judicial explication of this exception, such situations as telling a spouse about a patient's heart condition or impending death, or telling the employer of a roofer that the patient is subject to blackouts would probably qualify.

What type of form should be used to authorize the release of personal medical information?

The US Privacy Commission discovered that often when an individual applies for a job, life or health insurance, credit or financial assistance, or services from the government, the individual is asked to relinquish certain medical information. Although this is necessary in many cases, the commission found that individuals are generally asked to sign open-ended or blanket authorizations with clauses such as one requiring the recipient to "furnish any and all information on request." [25]

The American Psychiatric Association takes the position that such blanket consent forms are unacceptable, since they do not provide the patient the usual informed-consent protections. [26] The commission agreed and made the following recommendations:

> Whenever an individual's authorization is required before a medical care provider may disclose information it collects or maintains about him, the medical care provider should not accept as valid any authorization which is not:

(a) in writing;
(b) signed by the individual on a date specified or by someone authorized in fact to act in his behalf;
(c) clear as to the fact that the medical care provider is among those either specifically named or generally designated by the individual as being authorized to disclose information about him;
(d) specific as to the nature of the information the individual is authorizing to be disclosed;

(e) specific as to the institutions or other persons to whom the individual is authorizing information to be disclosed;

(f) specific as to the purpose(s) for which the information may be used by any of the parties named in (e) both at the time of the disclosure and at any time in the future;

(g) specific as to its expiration date, which should be for a reasonable time not to exceed one year.

Patients or former patients should not sign release forms that do not meet these criteria. Likewise, health care providers should refuse to honor requests that are not at least this specific on the grounds that the patient probably did not understand what he or she was consenting to when the patient signed the form. Health care providers should be obligated to contact the affected individual directly if they are suspicious of the quality of the consent and, thus, the legality of the release form.

Under what circumstances do health care professionals have the right to discuss the patient with other health care professionals?

Health care professionals who are *directly* involved in the patient's care may (and should!) discuss the patient's case among themselves without breaching confidentiality. However, they may not discuss the patient with other professionals who are not directly involved in the patient's care without consent *unless* the patient's identity is protected. Hypothetical cases (based on real cases but without names or other identifiers) can be openly discussed so long as the individual patient cannot be identified.

A related question is presented when a patient's case is presented at grand rounds at a teaching hospital. The patient is helped by the presentation, and it can be part of the patient's direct care. Nonetheless, as a matter of courtesy the patient's consent should be solicited prior to such "public" disclosure in grand rounds if the patient is easily identifiable and might reasonably object. Soliciting consent is unnecessary if reasonable steps are taken to protect the patient's identity.

Do health professionals have the right to discuss the patient's case with the patient's family without the express consent of the patient?

Only in general terms, such as "satisfactory" or "stable." No disclosures concerning diagnosis or prognosis to spouses or relatives should be made without the patient's express consent, if the patient is competent. If the family is making decisions for an incompetent patient, however, the family members have a right to all the information needed to give informed consent on the patient's behalf.

What recommendations did the Privacy Commission make concerning medical records?

The Privacy Protection Study Commission, established by Congress in 1974 to study privacy rights and record-keeping practices, issued its final report in 1977. This report is still the most complete and authoritative on the subject of the privacy of records. The commission found that medical records contain more information and are available to more users than ever before; that the control of health care providers over these records has been greatly diluted; that restoration of this control is not possible; that voluntary patient consent to disclosure is generally illusory; that patients' access to their records is rare; and that there are steps that can be taken to improve the quality of records, to enhance patients' awareness of their content, and to control their disclosure. Some of the commission's major recommendations are that:

1. Each state enact a statute creating individual rights of access to, and correction of, medical records, and an enforceable expectation of confidentiality for medical records.
2. Federal and state penal codes be amended to make it a criminal offense for any individual knowingly to request or obtain medical record information from a medical care provider under false pretenses or through deception.
3. Upon request, an individual who is the subject of a medical record maintained by a medical care provider, or another responsible person designated by the individual, be allowed to have access to that medical record, includ-

ing the opportunity to see and copy it; and have the opportunity to correct or amend the record.

4. Each medical care provider be required to take affirmative measures to assure that the medical records it maintains are made available only to authorized recipients and on a "need-to-know" basis.

5. Any disclosure of medical record information by a medical care provider be limited only to information necessary to accomplish the purpose for which the disclosure is made.

6. Each medical care provider be required to notify an individual on whom it maintains a medical record of the disclosures that may be made of information in the record without the individual's express authorization.[27]

Can patients be photographed by health care professionals?

Yes, but only with the patient's consent. Although most of the cases regarding photographs of patients involve the publication of the photos in newspapers or magazines, there are cases that involve physicians as well. Documentaries and television programs have also used film shot in hospitals and mental institutions. For example, in the late 1960s, filmmaker and lawyer Frederick Wiseman shot eighty thousand feet of film inside Bridgewater, a state mental institution in Massachusetts; the result was the documentary *Titicut Follies*. The Supreme Judicial Court of Massachusetts prohibited showing the film to the public in the state:

> The Commissioner and Superintendent, under reasonable standards of custodial conduct, could hardly permit merely curious members of the public access to Bridgewater to view directly many activities of the type shown in the film. We think it equally inconsistent with their custodial duties to permit the general public (as opposed to members of groups with a legitimate, significant interest) to view films showing inmates naked or exhibiting painful aspects of mental disease.[28]

Elsewhere in the decision, the court used the following words to describe the invasion of privacy: "collective, indecent intrusion";

"massive, unrestrained"; and "embarrassing." The patient's right to privacy is superior to any private, and usually superior to any public, interest in obtaining information about their medical care. This means that *no* film, not even photographs for the medical record, should be taken of patients without their prior consent.

This point was poignantly made in a 1976 case from Maine, which involved a patient who was dying of cancer of the larynx.[29] Both a laryngectomy and a subsequent radical neck dissection were performed. The surgeon took photographs, solely for use in the medical record and not for publication, of the progress of the disease. On the day before the patient died, the physician entered his room, placed some blue toweling under his head for color contrast, and took several final photographs. There was evidence that the patient raised a clenched fist and moved his head in an attempt to get out of the camera's range. His wife had also informed the physician she "didn't think that Henry wanted his picture taken." The trial court granted a motion for a directed verdict in favor of the physician, but the Maine Supreme Court reversed, saying that the physician's actions amounted to a violation of the patient's right to privacy:

> Absent express consent...the touching of the patient in the manner described by the evidence in this case would constitute assault and battery if it was part of an undertaking which, in legal effect, was an invasion of plaintiff's "right to be let alone."
>
> We are urged to declare as a matter of law that it is the physician's right to complete the photographic record by capturing on film B.'s appearance in his final dying hours, even without the patient's consent or over his objections. This we are unwilling to do.
>
> The facial characteristics or peculiar cast of one's features, whether normal or distorted, belong to the individual and may not be reproduced without his permission.[30]

What are some other examples of rights encompassed in the "right to privacy?"

Patients in a hospital have the following rights based on the right to privacy as enunciated in the cases discussed and referenced in this chapter:

1. To refuse to see any or all visitors
2. To refuse to see anyone not officially connected with the hospital
3. To refuse to see persons officially connected with the hospital who are not directly involved in the patient's care and treatment
4. To refuse to see social workers and chaplains and others not directly involved in the patient's care and to forbid them to view the patient's records
5. To wear his or her own bedclothes, so long as they do not interfere with the patient's treatment
6. To wear religious medals
7. To have a person of the patient's own sex present during a physical examination by a medical professional of the opposite sex
8. Not to remain disrobed any longer than is necessary for accomplishing the medical purpose for which the patient is asked to disrobe
9. Not to have the patient's case discussed openly in the hospital
10. To have the patient's medical records read only by those directly involved in treatment or the monitoring of its quality
11. To insist on being transferred to another room if the person sharing it with the patient will not let the patient alone or is disturbing the patient unreasonably by smoking or other actions.

NOTES

1. Ervin, *Civilized Man's Most Valued Right*, 2 Prism 15 (June 1974); cf. A. Westin & M. Baher, *Data Banks in A Free Society* (New York: Quadrangle, 1973), at 17–20; A. Miller, *The Assault on Privacy* (New York: New American Library, 1972), at 148–220. *And see generally* ABA Forum Committee on Health Law, *A Practical Guide to Access, Disclosure and Legal Requirement Relating to Hospital, Patient, Medical Staff and Employee Records* (Chicago: American Bar Association, 1987).

2. *Hammonds v. Aetna Cas. & Sur. Co.*, 243 F. Supp. 793, 801 (ND Ohio 1965).
3. Altman, "Physician-Patient Confidentiality Slips Away," *New York Times,* Sept. 27, 1983, at Cl.
4. Siegler, *Confidentiality in Medicine— A Decrepit Concept*, 307 New Eng. J. Med. 1519 (1982).
5. Other reasons supporting patient access are discussed in the preceding chapter, "Medical Records."
6. In *Curry v. Corn*, 277 N.Y.S. 2d 470 (1966), for example, the physician disclosed information to his patient's husband, who was contemplating a divorce action. In *Schaffer v. Spicer*, 215 N.W.2d 134 (S. D. 1974), the wife's psychiatrist disclosed information to the husband's attorney to aid him in a child custody case. Representative of the insurance cases are *Hague v. Williams*, 37 NJ 328, 181 A.2d 345 (1962), where the pediatrician of an infant informed a life insurance company of a congenital heart defect that he had not informed the child's parents of, and *Hammonds v. Aetna (supra* note 2), where the physician revealed information to an insurance company when the insurance company falsely represented to him that his patient was suing him for malpractice. Cases involving reporting to employers include *Beatty v. Baston*, 13 Ohio L. Abs. 481 (Ohio App. 1932), where the physician revealed to a patient's employer during a workman's compensation action that the patient had venereal disease; *Clark v. Geraci*, 208 N.Y.S.2d 564 (Sup. Ct. 1960), where a civilian employee of the United States Air Force asked his doctor to make an incomplete disclosure to his employer to explain absences, but the doctor made a complete disclosure including the patient's alcoholism; *Horne v. Patton*, 291 Ala. 701, 287 So. 2d 824 (1973), which involved the disclosure of a long-standing nervous condition; and *Alberts v. Devine*, 395 Mass. 59, 479 N.E.2d 113 (1985), *cert. denied*, 447 U.S. 1013 (1985), which involved disclosure of psychiatric information to a minister's clerical superiors that resulted in his not being reappointed. *And see* Note, *Breach of Confidence: An Emerging Tort*, Colum. L. Rev. 1426 (1982); and Winslade, *Confidentiality of Medical Records*, 3 J. Legal Medicine 497 (1982).
7. The most important exceptions are:
    1. Communications made to a doctor when no doctor–patient relationship exists
    2. Communications made to a doctor that are not for the purposes of diagnosis and treatment or are not necessary to the purposes of diagnosis and treatment (for example, who inflicted the gunshot wound and why)
    3. In actions involving commitment proceedings, wills, and insurance policies
    4. In actions in which the patient brings his physical or mental condi-

tion into question (for example, a personal injury suit for damages, raising an insanity defense, malpractice action against a doctor or hospital)

5. Reports required by state statutes (for example, gunshot wounds, acute poisoning, child abuse, motor vehicle accidents, and, in some states, venereal disease)

6. Information given to the doctor in the presence of another not related professionally to the doctor or known by the patient.

   And *see generally* Benesch & Homisak, "The Physician-Patient Relationship: Privileges and Confidentiality" in Zaremski & Goldstein, eds., *Medical Hospital Malpractice Liability* (New York: Callaghan & Co., 1987)

8. A. Westin, *Privacy and Freedom* (New York: Atheneum, 1967), at 7.

9. *Horne v. Patton, supra* note 6, 287 So. gd at 830.

10. Miller, *Personal Privacy in the Computer Age,* 67 Mich. L. Rev. 1091, 1107 (1968).

11. In 1939, for example, *Time* magazine published a story in its "Medicine" section with a photograph of the patient, a young woman who was receiving treatment for uncontrollable gluttony apparently induced by a condition of the pancreas (*Barber v. Time, Inc.,* 348 Mo. 1199, 159 S.W.2d 291 [1942]). Other cases, like *Horne v. Patton* (*supra* note 6), however, indicate that publication is not necessary to sustain an invasion-of-privacy action, for example, having unauthorized persons in a delivery room (*DeMay v. Roberts,* 46 Mich. 160, 9 W. 146 [1881]). Permitting unauthorized persons to view confidential medical records may also be an invasion of privacy.

12. Cal. Health & Safety Code sec. 2554 (Supp. 1970).

13. *Landeros v. Flood,* 17 Cal. 3d 399, 131 Cal. Rptr. 69, 551 P.2d 389 (1976) (physician could be liable to child for injury suffered from child abuse after physician's failure to report earlier incident of child abuse).

14. N.Y. Penal Code sec. 265.25 (Supp. 1969). And see generally Rose, *Pathology Reports and Autopsy Protocols: Confidentiality, Privilege and Accessibiliy,* 57 Am. J. Crim. Proc. 144 (1972); *and Denver Pub. Co. v. Dreyfus,* 184 Colo. 288, 520 P.2d 104 (1974) (autopsy reports open to public under Open Records Act). No statute is needed to authorize release of confidential medical information when a danger to the public exists. The leading case enunciating this exception, *Simonsen v. Swenson,* 104 Neb. 994, 177 N.W. 831, was decided by the Supreme Court of Nebraska in 1920. In that case, a man who was visiting a small town was seen by a physician who was also the physician for the hotel in which he was staying. The physician diagnosed syphilis and advised the patient to "get out of town," or he would tell the hotel's owner. When the patient remained in town, the doctor notified the landlady, who disinfected his room and placed his belongings in the hallway. The court decided that the doctor had the right to

reveal only as much information concerning a contagious disease as was necessary for others to take proper precautions against becoming infected, and that his actions under the circumstances were justified.

15. *E.g., Dennie v. University of Pittsburgh School of Medicine*, 638 F. Supp. 1005 (W.D. Pa. 1986); and *Commonwealth v. Petrino*, 480 A.2d 1160 (Pa. 1984), *cert. denied.* 471 U.S. 1069 (1985).

16. *Supra* note 7; And *see, e.g.*, Turkington, *Legal Protection for the Confidentiality of Health Care Information in Pennsylvania*, 32 Vill. L. Rev. 259, 302-72 (1987).

17. *Hofmann v. Blackmon*, 241 So. 2d 752 (Fla. App. 1970); and see *Wojcik v. Aluminum Co. of America*, 183 N.Y. S. 351, 357 (1959); and *Jones v. Stanko*, 118 Ohio St. 147, 160 N.E. 456 (1928).

18. *Davis v. Rodman*, 147 Ark. 385, 227 S.W. 612 (1921) (typhoid); Skillings v. Allen, 143 Minn. 323, 173 N.W. 663 (1919) (scarlet fever).

19. *Tarasoff v. Regents of U. of California*, 131 Cal. Rptr. 14, 551 P.2d 334 (1976).

20. Stone, T*he Tarasoff Decisions: Suing Psychotherapists to Safeguard Society*, 90 Harv. L. Rev. 358 (1976); and *see* Note, *Where the Public Peril Begins: A Survey of Psychotherapists to Determine the Effects of Tarasoff*, 31 Stan. L. Rev. 165 (1978).

21. *See, e.g., Mclntosh v. Milano*, 168 NJ Super. 466, 403 A.2d 500 (1979); *Bradley Center v. Wessner*, 287 S.E.2d 716 (Ga. Ct. App. 1982); *Williams v. US*, 450 F. Supp. 1040 (D.S.D. 1978); *Bardoni v. Kim*, 390 N.W.2d 218 (Mich. 1986); *Davis v. Lhim*, 355 N.W.2d 481 (Mich. 1983); Lipan v. Sears & Roebuck, 497 F. Supp. 185 (Neb. 1980); *duty to warn not found: Furi v. Spring Grove State Hospital*, 454 A.2d 414 (Md. 1983); *Cooke v. Berlin*, 735 P.2d 830 (Ariz. 1987); *Hinkelman v. Borgess Medical Center*, 403 N.W.2d 547 (Mich. 1987). *See generally* Note, *Psychiatrist's Liability to Third Parties for Harmful Acts Committed by Dangerous Patients*, 64 N.C.L. Rev. 1534 (1986).

22. *See* Barron, "A Debate Over Disclosure to Partner of AIDS Patients," New York Times, May 8, 1988, at E8; and Wilkes & Shuehman, "Holy Secrets," *New York Times Magazine*, Oct. 1988, at 57

23. See Siegler, *supra* note 4.

24. *See* discussion of "blanket consent forms" in ch. VI, at p. 93.

25. Privacy Protection Study Commission, *Personal Privacy in an Information Society* (Washington, DC: Government Printing Office, 1977), at 314. For example, most life insurance companies require all who apply for insurance to agree to release their medical records to the Medical Information Bureau, an information exchange agency operated by approximately 700 life insurance companies in the United States and Canada. Although the MIB has taken steps to try to maintain confidentiality of this information from nonmember companies and others, it still has an anachronistic

policy of refusing to release the medical information it maintains on an individual to the individual himself and, instead, requires the individual to designate a physician to receive the information "so that the nature and importance of the medical findings can be properly interpreted." *See* Entmacher, *Medial Information Bureau,* 233 JAMA 1370, 1372 (1975) (information on an individual's file can be obtained by writing MIB at Box 105, Essex Station, Boston, Mass. 02112).

26. *Id.,* citing American Psychiatric Association, *Confidentiality and Third Parties* (Washington, DC: APA, 1975), at 13.
27. *Personal Privacy, supra* note 25, at 293-314.
28. *Commonwealth v. Wiseman,* 356 Mass. 251, 259, 249 N.E.2d 610, 616 (1969).
29. *Berthiaume v. Pratt,* 365 A.2d 792 (Me. 1976).
30. *Id.* at 796–97. *See also Knight v. Penobscot Bay Medical Center,* A.2d 915 (Me. 1980). (A jury verdict in favor of the hospital in which the husband of a staff nurse had viewed a birth was affirmed. The jury had been instructed to find in favor of the plaintiff only if it found the intrusion was *intentional* and "would be highly offensive to a reasonable person.")

XII

Care of the Dying

Discussion of death has been transformed from taboo to high fashion. In the past two decades, there has been an avalanche of books, newspaper and magazine articles, movies, and Broadway plays on the subject. And the AIDS epidemic has made the discussion of terminal illness almost commonplace. Almost two million Americans die annually, and an additional million have a terminal diagnosis at any given time. Eighty percent die in hospitals or nursing homes. Thus, millions of people at any time are either dying or caring for dying relatives. Nonetheless, we remain a death-denying culture and continue to act as if immortality is attainable. Dying patients often know better, and even more than death, they fear isolation and pain.

It is also ironic that the most discussed patient right has been the "right to die," since death is the last thing most patients want, and most people rank access to health care well ahead of cessation of medical treatment. Nonetheless, the mere existence of new medical technologies has been seen by some to create a new obligation on the part of citizens to submit to their use, making the right to refuse treatment one of increasing importance. This chapter focuses on rights most pertinent to the dying patient and on the problems faced by families and professionals charged with their care.

Does a patient have a right to know that health care providers consider the patient terminal?

Yes. There is no justification for withholding this information from a patient. Surveys from the 1950s and early 1960s indicated that although 90 percent of all patients wanted to know their diagnosis, even if terminal, 60 to 90 percent of physicians preferred to withhold a terminal diagnosis.[1] Recent surveys reveal a dramatic shift, with 97 percent of physicians now indicating that their general policy is to disclose a terminal diagnosis.[2] Much of this shift can be attributed to society's changing attitudes and to more open discussion of death. It also reflects a more realistic assessment of who is dying, who must make plans for death, and thus, who has a right to know this information.

A few physicians still rationalize their preference for withholding a terminal diagnosis on the basis that such a decision can only be made by considering individual circumstances, and that it is seldom possible to know for certain that a patient is dying. But such a stance is easily translated into a policy of almost never telling, especially if the patient has a family onto which the physician can unburden his knowledge. And it fails to appreciate that what is ultimately at issue is truthfulness about uncertainty. A second rationale often advanced is that "the patient knows anyway." If this is true, not telling only forces both the patient and those around the patient to live the lie, increasing the patient's conflict and anxiety. A final argument is that the patient should not be forced to abandon hope. Hope may be essential, but it is not true that the only rational hope a person can harbor is that he will never die. A cancer patient may begin by hoping he does not have cancer, for example, then hope it will not be too painful, and finally hope he will live to see his daughter graduate from college. Honesty about a terminal diagnosis is not the same thing as denying the patient all hope.

To help the patient accept and deal with the diagnosis, the physician must, however, continue to talk with the patient openly and assure the patient that all that is medically reasonable will be done to keep the patient comfortable. These requirements have led some physicians to conclude that they do not have time for such "treatment," and therefore either the chaplain should talk with the patient about death, or no one should. This is as unique in medical practice as it is unjustified. In no other instance would the physician routinely share such confidential information with a third party. Giving the task of transmitting the prognosis to the chaplain is the symbolic way a physician "hands over" the patient, saying in effect, "There is no longer anything I can do for this patient."

Right implies duty, but it also implies option. If a patient clearly expresses a desire not to be told a terminal diagnosis, it is proper for the physician to ask whom the patient would like told the diagnosis and for the physician to inform that person instead. This should be rare, however. Death is inevitable, and planning for it is a personal responsibility we should not shirk off on our relatives and loved ones.

Does the dying patient have a right to demand that no one in the patient's family be told of a terminal diagnosis?

Yes. Even physicians who generally prefer not to tell dying patients about their condition often feel compelled to tell someone, usually the family. When this is done, the patient is deprived not only of the right to know the truth, but also of the rights of confidentiality and privacy. The patient has the right to make this demand and to expect physicians and nurses to respect it. A civil damage suit for breach of confidentiality would probably be the main legal action a patient could take against a doctor or nurse who refused to respect such a demand.[3]

One possible explanation for physicians telling families instead of patients is that once the patient is labeled terminal, the physician ceases to treat the individual alone and begins to treat the family unit as "the patient." Reluctance to candidly discuss diagnosis with the patient makes concentration on the family almost inevitable. The tragedy is the resulting isolation of the patient from truth and family. Both patient and family try to pretend all is well, although each knows better. Visits are uncomfortable for family members, who talk of trivialities like the weather or football, and almost unbearable for the patient whose impending death is transformed into a mockery.

When patients are denied the opportunity to discuss their own deaths, they are simultaneously stripped of their dignity as adults and are treated like children. Leo Tolstoy describes the dehumanizing effect in *The Death of Ivan Ilyich.*

> What tormented Ivan Ilyich most was the deception, the lie, which for some reason they all accepted, that he was not dying but was simply ill, and that he only need keep quiet and undergo a treatment and then something very good would result....This deception tortured him—their not wishing him to admit what they all knew and what he knew.... Those lies—lies enacted over him on the eve of his death and destined to degrade this awful, solemn act to the level of their visiting, their curtains, their sturgeon for dinner—were a terrible agony for Ivan Ilyich.

The "survivor knows best" attitude is illustrated by the words of a woman who described the death of her uncle as beautiful: "John died happy, never even realizing he was seriously ill."

Can a terminally ill patient give informed consent to treament if the patient does not know that the doctor considers the patient's condition terminal?

No. If the patient lacks this vital piece of information, the patient's probable motive for consenting to any medical procedure, the belief that it will help restore health, is based on misinformation. Consent given without the knowledge of one's terminal diagnosis is based upon false and misleading information, and therefore is not valid consent.[4]

Rationales for treating without telling vary from "if I don't do something, she'll turn to quacks" to "she's going to die soon, anyway, so she won't be able to sue me" to "I will perform this experimental treatment for the good of society and the advancement of medicine even though I know it will not help her." Such lines of reasoning are predictable. When a physician determines "I have done all I can for you," the physician's perception may change to "now, what can you do for me?" The patient loses all rights as a person in this instance, becoming little more than a passive object used as a "sample" in a medical experiment. Worse than being told nothing, the patient may be told that a proposed experimental treatment will help.[5]

The legacy of deception and withholding information from patients is that many patients no longer believe what health professionals tell them. Some oncologists, for example, report that the hardest task they have is persuading patients that they do have cancer. It is important for all of us that these attitudes, and the practices that breed them, be challenged and changed.

Can a competent adult patient refuse treatment even if the patient will die without it?

Yes. It is almost incredible that anyone could ever think that it was acceptable to force unwanted treatment on a patient. Competent adults may refuse *any treatment*, including lifesaving and life-sustaining treatment, and artificial nutrition and hydration. Procedures do not have rights, patients do. And the patient has the right not to have his or her body invaded by *any* medical procedure. The decision to undergo treatment is not a medical decision. It is a personal decision that can be legitimately made only by the patient who will be directly affected. If a person is empowered by law to decide to undergo medical treatment, it follows that the person is also empowered to decline such treatment. If a person cannot decline treatment, the "right" to decide whether or not to undergo a treatment becomes a sham, equivalent to a "right to agree with your doctor."

Mr. Quackenbush, a seventy-two-year-old man who refused to have his gangrenous legs amputated, provides an example. "His conversation did wander occasionally but to no greater extent than would be expected of a seventy-two-year-old man in his circumstances." He had shunned medical treatment for the past forty years. He was neither terminally ill nor comatose. If he had the amputations, he would live indefinitely, and if he did not, he would die. A New Jersey court explicitly concluded that the state's interest in preservation of life was not sufficiently compelling to override this patient's right to decide his own future regardless of the dim prognosis.[6] Other courts, including those in Arizona, California, Florida, Maine, Massachusetts, and Rhode Island have reached the same conclusion using similar reasoning.[7]

Contemporary courts have indicated that for competent patients, the finding of a "good prognosis" is insufficient to justify the state to forcibly treat these individuals. Whether a prognosis is "good" is not a medical issue and not an issue that can be resolved by an objective test. A prognosis is good or bad based on a subjective evaluation of the situation by the patient. Mr. Quackenbush's life would have been extended if his legs had been amputated. This may seem like a "good" prognosis to physicians or judges, but it did not seem like a "good" prognosis or good "quality of life" to Mr. Quackenbush.

It is seldom, if ever, proper for a state to force its view of a "good" life on a competent patient. The right to refuse medical treatment is not conditioned on the state's (or doctor's) finding, or not finding, that the proposed treatment is "good." Rather, it is based on the right of each citizen to make important personal health care decisions without interference by the state.[8] The tragedy of modern American medicine is that we continue to unlawfully force people to endure invasive and expensive treatment they do not want while simultaneously denying others treatment they desperately want and need.

What does it mean to be mentally competent?

In the context of medical care, a person is competent if the person can understand the nature and consequences of the illness or condition and the nature and consequences of the proposed medical procedure. This is a factual question, requiring a discussion with the patient. If an individual understands the information needed to give informed consent, then that individual is competent to either give or refuse consent.[9] An unconscious patient is obviously incompetent, as are young children. With these exceptions, no simple categorization of incompetence exists.

A patient can be rendered incompetent by mental illness, mental retardation, senility, drugs, alcohol, pain, and other conditions. But the fact that someone is mentally retarded or committed to a mental hospital does not automatically render the person incompetent. Ms. Yetter, for example, was a sixty-year-old woman who had been involuntarily committed to a mental hospital with a diagnosis of schizophrenia. When a lump in her breast was discovered, she refused to consent to a biopsy, which would be followed by removal of her breast should cancer be found. The refusal was based on the fact that her aunt had died after a similar procedure, and she was afraid of the operation. At the court hearing, Ms. Yetter testified that she did not want to undergo the procedure because "it could interfere with her genital system, affecting her ability to have babies, and would prohibit a movie career." The court found that, even though she was getting more delusional, she had consistently opposed the surgery during her lucid periods and therefore was competent to refuse the surgery. The court specifically concluded that the patient's refusal to undergo the procedure was not caused by her delusions, and that she knew that she might die as a result of her refusal.[10]

Another case involved Ms. Candura, a seventy-seven-year-old woman who suffered from gangrene in her foot and refused to undergo a recommended amputation. The court found that the patient was combative and defensive at times, that she was sometimes confused, that her mind wandered, and that her concept of time was distorted. Nonetheless, she demonstrated a high degree of awareness and acuity. The patient made clear she did not want the operation, even though she knew her decision probably would shortly lead to her death. The court concluded that she had made her choice "with full appreciation of the consequence." As a result, she was found competent.[11]

The *Candura* case also illustrates another important fact: A patient's competence is often not questioned as long as the patient agrees to undergo the proposed medical procedure. As the court pointed out: "Until she changed her original decision and withdrew her consent to the amputation, her competence was not questioned. But the irrationality of her decision does not justify a conclusion that Mrs. Candura is incompetent in the legal sense. The law protects her right to make her own decision to accept or reject treatment, whether that decision is wise or unwise." [12]

Thus, if physicians determine that a patient is competent to consent to an operation, courts will use that conclusion to determine the patient competent to refuse consent as well. In fact, in the *Candura* case, the physicians were willing to perform the procedure on the basis of her consent alone, even at the time of the trial.

All adults are presumed competent until proven otherwise. Therefore, anyone who wishes to treat a person without consent (on the basis of that person's incompetency) should be required to prove that because of some mental or physical disease or disability, the patient cannot understand the nature and consequences of the proposed procedure and the consequences of refusing to undergo the procedure. The refusal to undergo the procedure, however "irrational" that decision may appear, is never alone sufficient to prove incompetence.

Can a patient's family determine if a patient is competent?

Since competence is a *factual* question, anyone who can determine the relevant facts can determine competence.[13] Of course, only a court can appoint a legal guardian to act on behalf of an incompetent patient. As previously noted, competence is almost never questioned except when a patient *refuses* a recommended treatment. This, of course, is illogical, since consent should not have been sought while the patient's competence was in question.[14]

What are the legal limits on the right of competent persons to refuse treatment?

There really are none. Nevertheless, courts have said that the individual's right to refuse treatment must be balanced against a variety of state interests if the state challenges the patient's decision. These state interests include (1) the preservation of life, (2) the protection of innocent third parties, (3) the prevention of suicide, and (4) maintenance of the ethical integrity of the medical profession.

1. *The preservation of life.* The clearest statement on the state's interest in preserving human life is by the Massachusetts Supreme Judicial Court: "The constitutional right of privacy...is an expression of the sanctity of individual free choice and self-determination as fundamental constituents of life. The value of life as so perceived is lessened not by a decision to refuse treatment, but by the failure to allow a competent human being the right of choice."[15] In other words, life without liberty is not the type of life the state should compel.[16]

2. *The protection of innocent third parties.* This is probably the strongest state interest that has been expressed by the courts. Usually it has been interpreted to mean that the state should be able to require treatment of a parent because the parent's death would have an adverse psychological and financial impact on the children. Although this argument has enormous emotional appeal, it has equally enormous conceptual problems. First, it sets up two classes of persons, parents and nonparents, and penalizes parents by depriving them of the right to refuse treatment because they are parents. It would also mean that courts would have to make determinations on a case-by-case basis of the impact of the death of a parent on the children. Would an impoverished child abuser have more of a right to refuse treatment than a wage-earning model parent? And should this same logic be used to reject divorce petitions or to prohibit parents from moving with their children to other states or countries? Finally, application of this concept would require a competent adult to undergo a medical procedure for the benefit of another person. This could present untold problems in the future if permitted by the courts.[17]

The most common way this interest is used is by lower courts to require Jehovah's Witnesses to have blood transfusions. What really seems to be at issue in these cases, however, is religious persecution. Many judges simply seem to believe that the Witnesses' belief that accepting blood will condemn them for eternity is irrational, and thus, it makes them *de facto* incompetent. Others have concluded that it is appropriate to order blood on the basis that no religious obligation is violated if the blood is forced on the Witness. This latter belief, however, is not shared by most Witnesses themselves.

3. *Prevention of suicide.* Although once seen as a problem, courts now relegate this concern to footnotes. Like murder, suicide (or self-murder) requires a specific intent to cause death and an action putting the death-producing agent in motion. In cases of refusing treatment, it is not that the patient wants to die, but rather that the patient does not want to undergo a certain treatment and is willing to accept death as a consequence of that decision. Additionally, the patient does not set the death-producing agent in motion, since patients do not cause the illnesses or conditions they will die from.

4. *The maintenance of the ethical integrity of the medical profession.* Medical ethics do not require the forcible treatment of unconsenting competent patients; therefore courts have not ordered treatment on this basis.[18] Courts have almost never ordered competent individuals to undergo any treatment more invasive than a blood transfusion and are unlikely to do so in the future. The specter of a competent patient being strapped down, chemically restrained, or even arrested, and forced to undergo a medical procedure he or she does not want is so abhorrent to a free society that all medical professionals (and lawyers and judges) should work to ensure it never occurs. As one judge has pointed out, "The notion that an individual exists for the good of the state is, of course, quite antithetical to our fundamental thesis that the role of the state is to ensure a maximum of individual freedom of choice and conduct."[19]

Who has the right to refuse lifesaving treatment for an incompetent patient?

This is one of the most controversial areas of medical jurisprudence. Courts have begun to realize that neither good medical practice nor the law requires that incompetent patients be continuously treated with all the resources that medical science and technology have to offer. Judges have concluded that competent individuals have the right to refuse treatment, and incompetent persons should not be deprived of this right simply because they are incapable of expressing their wishes. Accordingly, a way must be found to enable incompetent individuals to have a similar right exercised on their behalf.

The most famous case, that of Karen Ann Quinlan, was decided by the New Jersey Supreme Court.[20] At the time the case was brought to court, Ms. Quinlan was twenty-one years old and in a "chronic vegetative state." She required constant medical attention, and it was thought that she would die if the respirator was removed. All the physicians who examined her concluded that she suffered from incurable brain damage and would never awaken from her coma- tose condition. Her parents wanted the respirator removed, but the physicians refused. Her father petitioned the court for appointment as her guardian with the explicit power to order the removal of the respirator. The trial court rejected his request.

On appeal, the New Jersey Supreme Court decided that the con- stitutional right to privacy was broad enough to include the right to refuse treatment in such a dire situation, and that this right could be exercised on the patients' behalf by a guardian. The court also found that physicians were leery of terminating treatment in such cases because of their fear of possible civil and criminal liability. Because of this, the court set up a procedure to try to make physicians more comfortable in such cases. Specifically, the court decided that if the family requested discontinuance of treatment, and if the patient's physician concluded that there was "no reasonable possibility" of the patient's returning to a "cognitive, sapient state," and if the "guard- ian and family" of the patient concurred, and if a hospital "ethics committee" (made up of physicians, lawyers, social workers, and theologians) agreed with this conclusion, then the participants in the decision would be immune from civil and criminal liability.

The second most important case was decided by the Massachu- setts Supreme Judicial Court. Joseph Saikewicz was a sixty-seven- year-old profoundly retarded resident of a state institution who was suffering from acute myeloblastic monocytic leukemia.[21] Expert medical testimony indicated that 30 to 50 percent of patients achieve remission with chemotherapy (usually for two to thirteen months), that persons over sixty have more difficulty tolerating the treatment than younger people, that success rates are lower in older people, that the side effects would make Mr. Saikewicz feel uncomfortable and sicker, and that he would probably have to be restrained to administer the treatment properly. Without the treatment, Mr. Saikewicz would die comfortably in a few weeks to several months. As a result of these factors, the *guardian ad litem* (a guardian appointed for duration of the lawsuit only) and the physicians recommended that chemotherapy not be administered. Both the trial court and Supreme Judicial Court agreed with this recommendation.

In almost every area, the *Saikewicz* court agreed with the *Quinlan* court. They disagreed in one respect, however. Unlike the New Jersey court, the *Saikewicz* court concluded that if the parties wanted immunity from civil and criminal liability, they would have to apply to a court (not merely a hospital ethics committee). It should be pointed out, however, that neither court required recourse to either courts or ethics committees. These bodies must be consulted only if immunity from suit is seen as necessary. Since no health care provider has even stood trial for terminating treatment of such seriously ill individuals, the need for immunity ranges from nonexistent to highly questionable. This point can be underscored—the fact is that the highest courts of the first two states that had the opportunity to decide the issue decided that it is *not* always in the best interests of the seriously ill incompetent patient to receive treatment, that physicians and families or guardians working together can make decisions to suspend treatment based on certain criteria, and that mechanisms exist, when desirable, to have the legality of these decisions reviewed before they are acted on.[22]

What criteria have courts established to determine when treatment of an incompetent person can be stopped or withheld?

Both the *Quinlan* and *Saikewicz* courts adopted the "substituted judgment" doctrine. The *Saikewicz* court stated that the primary test is "to determine with as much accuracy as possible the wants and needs of the individual involved." The *Quinlan* court stated that "the only practical way to prevent destruction of the right [to refuse treatment] is to permit the guardian and family of Karen to render their best judgment...as to whether she would exercise it in the circumstances." The goal is to treat the incompetent person as the person would choose to be treated if the person were competent.

In 1990, the United States Supreme court decided that the United States Constitution does not require states to adopt any particular decision making rule for incompetent patients.[22] In the case of Nancy Cruzan, a young woman in exactly the same physical condition as Karen Ann Quinlan had been (except that Ms. Cruzan required only tube feeding to continue to live), the Missouri Supreme Court had ruled that only Nancy herself could give an informed refusal of life-sustaining medical treatment. Thus, unless her family could prove by "clear and convincing evidence" that Nancy herself had said or written that she would never want such treatment under these circumstances, the state of Missouri could require treatment to be continued indefinitely. The United States Supreme Court affirmed this decision, saying that nothing in the Constitution prohibited Missouri from adopting such a rigid rule because it was reasonably related to a legitimate state purpose of protecting the lives of incompetent patients who (unlike Nancy Cruzan) did not have loving families to look out for their interests. Although the *Cruzan* opinion did not change existing law and procedures in any state, it does mean that state courts and state legislatures now have broad powers (consistent with their own state constitutions) to regulate the procedures for withholding and withdrawal of life-sustaining treatment from incompetent patients. It also means that living wills and durable powers of attorney will become more prominent in American medicine.

How can we make sure that we will be treated the way we want to be treated should we become incompetent or no longer able to participate in decision making?

There is no way to be certain. The only certain thing is that when we become comatose or incompetent, others will have to make treatment decisions for us. If they know what decisions we would make, we can hope and trust that they will honor our wishes. The law supports them if they do. But to follow our wishes, they must know what they are. It is therefore critical that each of us tell our relatives and friends how we want to be cared for in the event of incompetence.

Many people have said simply, "I don't want to be like Karen Quinlan," meaning that they did not want to be kept alive on a mechanical ventilator if they were in a permanent coma. This type of statement is extremely helpful to physicians and families. But thankfully, most of us will not end up like Karen Quinlan. *The best strategy is to tell our physician and family how we want to be treated in various situations and, in addition, to designate one or more persons to act on our behalf to make treatment decisions when we are unable to make them ourself.* This can be done verbally, but it is best to write down directions in a form that can be understood by others (like a living will) and to formally designate another person (a friend or relative) to make decisions on our behalf through a document called a "durable power of attorney." An example of such a document is set forth in appendix C.

What is a living will?

The term "living will" (also termed an "advance directive") was coined by Luis Kutner in 1969; it describes a document in which a competent adult sets forth his or her wishes concerning medical treatment in the event he or she becomes incapacitated in the future.[23] In this sense, it is like a "will," but since it takes effect prior to death, it is termed a "living" will. Public interest is intense in this mechanism. Because of the absence of specific judicial sanction and the lack of clear rules regarding their execution and use, many individuals and organizations have long advocated that states pass specific statutes supporting the living will.

California enacted the first living will statute in 1976, designating it a "Natural Death Act," a term that many other states have used as well. California should receive considerable credit for enacting the first statute, but the price was very high. The statute is *extremely* narrow. A "binding" declaration can only be executed fourteen days or more *after* the declarant has been diagnosed as suffering from a terminal illness, making the person a "qualified patient." In order to qualify as "terminal," death must be "imminent," i.e., the patient must be dying soon *whether or not* life-sustaining procedures are used. As the President's Commission for the Study of Ethical Problems in Medicine noted, the fourteen-day waiting period under such circumstances requires "a miraculous cure, a misdiagnosis, or a very loose interpretation of the word 'imminent' in order for the directive to be of any use to the patient." [24] Even though the California statute was inspired in part by the *Quinlan* case, the statute does not apply to cases like hers because Ms. Quinlan was not terminally ill, and her death was not imminent.

Most states have similar limitations on the individuals covered, generally denoting them "terminally ill." This removes from their protection the very categories of patients who are likely to need this protection the most, patients like Earle Spring (kidney dialysis), William Bartling (mechanical ventilator), and Claire Conroy (tube feeding). As the President's Commission also noted, "such a limitation greatly reduces an act's potential." [25]

By 1989, more than 40 states had enacted legislation. Many of the statutes are different and none is ideal. All include specific instructions that must be followed in the execution of an advance directive, and most set out a model form for the directive in the statute itself. Some contain California's "imminent" dying language, and some permit the declarant to refuse only "artificial life-sustaining procedures."

What are the major problems with current living will statutes?

Almost all living will statutes suffer from the following shortcomings:

1. They are restricted to the terminally ill and thus exclude from their protection the vast majority of elderly individuals; the term, "terminally ill" is so vague that it is subject to arbitrary interpretation and application.
2. They limit the types of treatment a person can refuse to those that are "artificial" or "extraordinary," thus excluding many burdensome treatments; and these vague terms lead to arbitrary interpretations.
3. They do not permit an individual to designate another person to act on the individual's behalf (like a durable power of attorney) and do not set forth criteria under which

the person so designated is to exercise this authority, thus greatly restricting the usefulness of the document in cases not precisely predicted by the individual.

4. They do not require health care providers to follow the patient's wishes as set forth in the declaration, thus the rights of the patient are not treated as superior to those of the health care providers.

5. They do not explicitly require health care providers to continue palliative care (comfort and pain relief) to a patient who refuses other medical interventions.

Because of these shortcomings, living will statutes cannot resolve the many complex issues discussed in this chapter. As presently drafted, what these statutes primarily do is say that if a patient is terminally ill, and if the physician can do nothing to sustain the patient's life, and if the patient does not want life sustained, and the doctor agrees with the patient's decision, then the doctor *may* (but does not have to) follow the patient's desire and be assured criminal and civil immunity.

One proposal to remedy these shortcomings is the Right to Refuse Treatment Act, a "second-generation" living will act developed by the Legal Advisers Committee of Concern for Dying and set forth in its entirety in appendix D. The President's Commission wrote:

> The Act enunciates a competent adult's right to refuse treatment and provides a mechanism by which competent people can both state how they wish to be treated in the event of incompetence and name another person to enforce those wishes. In terms of its treatment of such central issues as the capacity to consent and the standard by which a proxy decision maker is to act, the Right to Refuse Treatment Act is carefully crafted and in conformity with the Commission's conclusions.[26]

What is a durable power of attorney?

A power of attorney is a written document wherein a person gives someone else the authority to perform certain acts on that person's behalf, consistent with that person's directions. The person given authority to act in a power-of-attorney document is an agent, and may be termed the attorney-in-fact. The person who executes such a document is called the principal. Ordinarily, powers of attorney cease to be effective when the principal becomes incompetent. *Durable* powers of attorney, however, continue in effect. The principal can also provide that the power "springs" into effect only upon the principal's incompetence (this is usually how durable powers of attorney for health care are drafted).[27]

Only a handful of states have enacted durable power-of-attorney statutes that pertain specifically to health care decisions. However, there is nothing in the durable power-of-attorney statutes in other states to prevent this document from being used to designate an individual to make health care decisions on one's behalf, and use of this document has been widely recommended for this purpose.[28]

Because none of us can accurately predict how we will die or what medical interventions might be available to prolong our dying, it is prudent for all of us to execute a durable power of attorney for health care (also called a health care proxy). In addition to not having to predict the future, this document names an individual to make decisions in our behalf, and gives the physician a person with whom to discuss treatment options. Living wills can still be used to discuss one's wishes with one's proxy, and can even be given to the proxy to help in decision making. But to avoid misinterpretations and potentially drawn out battles over what you "really meant" in your living will, it seems best that only the proxy be given this document. Examples of both a health care proxy alone, and a health care proxy combined with a living will, appear in appendix C. The forms can be modified to reflect the individual's own wishes, and in states that have specific legal requirements for durable powers of attorney for health care, will have to be modified to comply with state law as well.

Can parents refuse potentially life-sustaining treatment for their children?

Usually not, even for religious reasons. As the US Supreme Court stated in another context, "Parents may be free to become martyrs themselves. But it does not follow they are free...to make martyrs of their children."[29]

Parents have a legal obligation to provide their children with "necessary medical care." When alternative modes of treatment are available, and each alternative is consistent with generally accepted medical practice, parents may choose among them. But when the only alternative is nontreatment, parents may lawfully choose this option only if it is consistent with the "best interests" of the child. To deny the child beneficial treatment can be child neglect, and the state has an obligation to exercise its *parens patriae* power to protect children from such neglect.

So important is society's view of the sanctity of human life that when a duty to treat exists and treatment is withheld, both the parents and the physician could be charged with homicide.[30] Only one such case has ever been brought in the United States, however. That case involved a charge of attempted homicide for the initial failure to treat Siamese twins born in Danville, Illinois, but it was dropped for lack of evidence.[31] The children were eventually successfully separated, and at last report, they were doing "moderately well." The best interests standard has a long legal pedigree, but is often difficult to apply.

What are the "Baby Doe" regulations?

A Down syndrome baby, known only as Baby Doe, died in Bloomington, Indiana, on April 15, 1982, at the age of six days, following a court-approved decision that routine lifesaving surgery be withheld. The infant had a tracheoesophageal fistula (a connection between the trachea and esophagus that makes eating impossible) that was not repaired; instead, the child was medicated with phenobarbital and morphine and allowed to starve to death. The court believed that if there was a dispute among physicians about treatment, the parents should be able to withhold treatment. Given existing legal principles, however, that require treatment if it is in the child's best interests, it seems that legally and ethically, Baby Doe should have been treated.[32] The public, accordingly, was properly outraged that he was not.

On the strength of the Baby Doe case, the Department of Health and Human Services (HHS) wrote a letter to approximately 7000 hospitals, on May 18, 1982, putting them on notice that it was unlawful (under section 504 of the Rehabilitation Act of 1973) for a recipient of federal financial assistance to withhold from a handicapped infant nutritional sustenance or medical or surgical treatment required to correct a life-threatening condition if (1) the withholding is based on the fact that the infant is handicapped, and (2) the handicap does not render treatment or nutritional sustenance contraindicated.

In emergency regulations published in March 1983, HHS required the conspicuous display of a sign containing the substance of the May 1982 letter in each delivery ward, maternity ward, pediatric ward, nursery, and intensive care nursery. Included in the notice was a toll-free, twenty-four-hour "hot-line" number that individals with knowledge of any handicapped infant being discriminatorily denied food or customary medical care were encouraged to call. The HHS officials were given authority to take "immediate remedial action" to protect the infant, and hospitals were required to provide access to their premises and medical records to agency investigators. These regulations, and their successors, were ultimately invalidated by the US Supreme Court,[33] but they served to focus public debate on the appropriate treatment of handicapped newborns and the respective roles of the state and federal governments in protecting children.

When is refusing treatment for a child considered child neglect?

The federal Child Abuse Amendments of 1984, among other things, explicitly brand the withholding and withdrawal of medically indicated treatment and nutrition from disabled infants as a type of child abuse. The law requires individual states, as a condition for continued federal funding of their child abuse programs, to establish special procedures to deal with this form of child abuse.

On April 15, 1985, HHS issued final regulations to implement the new law. Treatment may be withheld if:

1. The infant is chronically and *irreversibly* comatose;
2. The provision of such *treatment would merely prolong dying*; not be effective in ameliorating or correcting all of the infant's life-threatening conditions; *or otherwise be futile* in terms of survival of infant; or
3. The provision of such *treatment would be virtually futile* in terms of the infant's survival and the treatment itself under such circumstances would be inhumane.[34] (emphasis added)

The "virtually futile" exception provides latitude for physicians to make reasonable medical judgments regarding most treatments. Medical judgment is given central authority and specifically includes the withholding of "other than appropriate nutrition, hydration, or medication" to an infant when any of the exceptions listed above apply. Whether or not one of the exceptions applies is determined solely on the basis of "reasonable medical judgment," which the regulation defines as a "medical judgment that would be made by a rationally prudent physician, knowledgeable about the case and the treatment possibilities with respect to the medical conditions involved." Nutrition and hydration can be withheld from an infant when they are not "appropriate" in the attending physician's "reasonable medical judgment."

The Baby Doe episode and the Child Abuse Amendments of 1984 have heightened our awareness of the rights of handicapped newborns. But the law now is almost precisely the same as it has been for the past two decades: Withholding necessary medical treatment can be child neglect if treatment would be in the child's best interests. Enforcement of this law is and should be the responsibility of the states, not the federal government.[35]

May the child's "quality of life" be taken into account in making a decision not to treat?

The answer to this question depends to a large extent on what is meant by the ambiguous term "quality of life."

A lower-court case in Maine involved a newborn who was blind, had no left ear, some abnormal vertebrae, and "some brain damage." The child also had a tracheoesophageal fistula, and the parents and physicians decided that this defect should not be corrected because the child would have a life "not worth preserving." The court concluded that "the doctor's qualitative evaluation of the value of life to be preserved is not legally within the scope of his expertise." Since the repair of the fistula was in no sense "heroic," and the treatment did not involve serious risk, the court said that the procedure must be performed.[36]

Courts will not permit nontreatment decisions to be made solely on the basis of the incompetent person's future mental or physical

handicap. In an extreme case, however, where the child will not
live long and will only experience pain or suffering, it may be in the
child's best interests not to be given specific treatments.

What is CPR (cardiopulmonary resuscitation) for?

The National Conference on CPR and ECC (electrocardiac con-
version), cosponsored by the American Heart Association and the
National Academy of Sciences, concluded: "The purpose of CPR is
the prevention of sudden, unexpected death. CPR is not indicated in
certain situations such as in cases of terminal irreversible illness where
death is not unexpected."[37]

CPR is contraindicated when it will do no good; that is, the patient will die soon anyway, and nothing can be done to stop the course
of the disease. When the patient's condition is hopeless, no medical
intervention is ethically or legally required because all medical
interventions are useless. Although there is room for debate about
the normative aspects of such a prognosis, there are cases where
such a decision is basically medical (such as those in which CPR is
known to be futile), and in these cases, the decision can and should
be made by the attending physicians.

The Massachusetts Court of Appeals seems to have believed it
had just such a case when it was asked to review the appropriateness
of a DNR ("do not resuscitate") order on Shirley Dinnerstein.[38] She
was sixty-seven years old and had been suffering from Alzheimer's
disease for six years. She was getting progressively worse. At the
time of the hearing, she was confined to a hospital bed in "an essentially vegetative state, immobile, speechless, unable to swallow without choking, and barely able to cough...her condition is hopeless...it
is difficult to predict exactly when she will die [but] her life expectancy is no more than a year." Her family, consisting of a son–
physician and a daughter who lived with her prior to her admission
to the nursing home, concurred in the DNR order.

The court viewed CPR as "highly intrusive" and "violent in nature," and Ms. Dinnerstein as a patient "in the terminal stages of an
unremitting, incurable mortal illness." The court concluded CPR
would do nothing for her and therefore was not a "significant treatment choice or election." Given her condition, which was exclu-

sively diagnosed by her physicians, the court concluded the DNR order was "a question peculiarly within the competence of the medical profession of what measures are appropriate to ease the imminent passing of an irreversibly, terminally ill patient in light of the patient's history and condition and the wishes of her family."[39]

The futility of a medical intervention is an issue peculiarly within the competence of the medical profession. And if CPR is futile, then obviously CPR is not indicated. Indeed, if CPR is known to be futile, and given that CPR is intrusive and unproductive, there seems no justification to inflict it on a patient regardless of the wishes of the family. The "hopeless" prognosis is more like the pronouncement of death than a decision about elective surgery, and the role of the family should be just as irrelevant in this type of DNR decision as it is in a determination-of-death decision.

Can a patient refuse CPR?

A patient can refuse any medical intervention, including CPR. CPR seems to have been treated differently from other medical procedures because it is administered under emergency conditions and, if successful, can be lifesaving. Physicians have a legal "privilege" to treat patients in emergencies without consent, because time is of the essence and obtaining informed consent under such circumstances is impossible. Nonetheless, if the emergency can be anticipated, and if the patient refuses consent to an intervention in advance, such as CPR or administration of a blood transfusion, the intervention cannot be either legally or ethically imposed on the nonconsenting patient.

Older physicians say that when blood transfusions were first introduced, no one could die in a hospital without getting a blood transfusion. Although CPR is almost twenty years old, the general rule today often seems to be that no one can die in a hospital without CPR. This is medical practice in many institutions, *but it is not the law*. Patients have the legal right to refuse to be resuscitated, and physicians have an obligation to discuss CPR with them if there is a reasonable possibility that it may be used on them during their hospital stay. CPR when done in a hospital is an extremely invasive procedure, involving the possibility of placing tubes in the trachea for artificial respiration, electrocardial shock, intravenous lines, cardiac medications, and possibly the placement of a temporary pacemaker. If a patient refuses it, the physician will enter a "do not resuscitate order" (DNR) in the patient's chart. This means that the patient will not be resuscitated in the event the patient's heart stops beating, and the patient will die. Patients and families must realize, however, that CPR is not magic. Unless the cardiac arrest is sudden and unexpected, even if the patient is resuscitated, he or she will likely die soon. The probability of CPR's success depends largely on the patient's medical condition.[40]

Can a DNR order be written
based on a patient's poor quality of life?

Only if the patient consents to this quality-of-life judgment, or the decision is consistent with the patient's prior wishes. For example, Sharon Siebert was forty-one years old, had been seriously brain-damaged in an operation five years previously and, according to her physicians, had a life expectancy of thirty-seven more years. She had the mental age of a two-year-old with no prospects for improvement. She had to be fed artificially because she could not swallow, was confined to a bed or wheelchair, and could communicate only "slightly" and "simply." A DNR order issued by her physician, with the consent of her parents, was challenged by a friend, Jane Hoyt. Following a court hearing, the DNR order was revoked on the basis that it was proper only if Ms. Siebert herself would have wanted it, and there was no evidence that she would have. The guardian "must consider all of the factors which his ward would consider were she able, and, so far as possible, evaluate them as she would."[41]

Can a patient demand CPR even if it is not medically indicated?

No. Patients do have a right to refuse any treatment, but do not have a right to demand that their physicians mistreat them by invading their bodies in ways that are contrary to good medical practice. For example, if CPR is *never* successful in hospitalized patients with metastatic cancer, it is useless and futile to use this technique in this category of patients[42]; therefore physicians should not do so, and patients have no right to demand that they do. Likewise, patients have no right to demand a kidney, heart, or liver transplant if such a transplant is not "medically indicated" and would be futile.

Depriving patients of futile treatments does not implicate autonomy. As a philosopher and a philosopher–physician properly noted:

> Patients or families who demand CPR when it will almost certainly be futile almost always do so because of denial of the prospect of death, magical thinking about the miracles of modern medicine, or benign television images of CPR. The physician who offers CPR under these conditions only supports these natural defense mechanisms and suggests, moreover—regardless of what he or she might say to the contrary—that CPR does offer something of value. Under these conditions, the patient's choice of CPR will be based on a misunderstanding of its benefits, and therefore will frustrate the pursuit of autonomy rather than serve it.[43]

It should be noted, however, that the state of New York has a 1988 law mandating the use of CPR unless the patient or the patient's proxy has refused it.[44] This law improperly treats CPR as a unique situation and seems to say that physicians must use it even when it is futile. It is likely that this law will be amended, repealed, or ignored in the area of futile resuscitation. Every JCAH-accredited hospital must have a DNR policy, and patients have a right to know the hospital's DNR policy and, of course, to refuse CPR in advance.

Do patients have the right to demand that their physician give them sufficient pain medication to relieve their pain?

Yes. Physicians have no right to require patients to suffer or to withhold needed and medically appropriate pain medications. Nonetheless, it is almost universally agreed among physicians that approximately 90 percent of terminally ill patients do not receive proper pain management. This is a major medical scandal. There is no law that prohibits physicians from giving patients all the pain medication needed to make them comfortable, even if such medication shortens the patient's life or makes a cardiac or respiratory arrest more likely. Patients have no legal or ethical obligation to suffer, and physicians *do* have a legal and ethical obligation to alleviate that suffering (assuming, of course, this is the wish of the patient).[45]

Do dying patients have the right
to use narcotic drugs for pain relief?

The general answer is that although there is no legal right, as a matter of public policy, even currently controlled and illegal narcotic drugs (and arguably psychedelic drugs) should be made freely available to dying patients. Experience demonstrates that, to be effective in pain relief, doses need not be so high as to distort reality. Notions that these drugs should be denied terminally ill patients because they might become addicted or their chromosomes might be damaged are almost always ridiculous on their face. Where drugs have been used, much success has been reported. When he was dying of cancer, columnist Stewart Alsop wrote eloquently of the experience and suggested that patients be allowed *"to decide for themselves"* how much painkilling drug they will take. It is, after all, they, not the doctors, who are suffering the agonies.[46] This is another example of a political "right" that patients can translate into a legal right or institutional policy only by exerting political pressure.

Do patients have the right to demand
that physicians end their lives?

No. The rights patients have in regard to treatment have to do primarily with refusing treatment and choosing among medically appropriate treatment alternatives. There have been proposals to enlarge the physician's role to that of helping terminally ill patients commit suicide, but these have so far been rejected.[47] Public opinion polls, for example, show that an overwhelming majority of the public (74 percent) and almost all physicians (88 percent) agree that patients in severe pain should have all the medications they need to ease the pain, even if this might shorten their lives. On the other hand, only a bare majority of the public (57 percent) and a minority of physicians (30 percent) think it would be right to actually end the life of a terminally ill patient in severe pain on request of the patient.[48] This issue is likely to be the subject of continued, heated debate inside and outside the medical profession.

Citizens have no obligation to seek medical attention even if their lives are in danger, and they have the right to commit suicide in private. Currently, however, they have no legal right to demand that their suicide be assisted by the medical profession. The primary argument in favor of such assistance is that the medical profession has the drugs available and is best able to administer them. The primary argument against is that the medical profession is built on trust in healing and alleviating suffering, and that adding the act of ending life, even with consent, would radically alter the role of physicians and erode the public's trust in them. The case is a very close one, and unless physicians begin to respond more humanely with pain medication at the end of life, public calls for legalizing medical assistance in death will inevitably increase and ultimately prevail.[49]

Do patients have the right to die at home?

The theoretical answer is yes. But, as a practical matter, society has made it extremely difficult for individuals to die at home. Although most would prefer to, less than 20 percent of all Americans die at home. To do so requires not only the strong resolve of the dying patient, but also the cooperation of the patient's family or the person or persons with whom the patient is living. Cooperation of the medical profession is also important, since without it, pain-relieving drugs and other medications cannot be obtained. It has been persuasively suggested that unless we develop a creative and compassionate program of home care for AIDS patients, the depersonalizing nature of institutional care, coupled with its high cost and the socially unpopular nature of the disease, will "leave individuals with no alternative to the indignities of their final days but to end them quickly."[50]

Are dying patients especially vulnerable to experimental and nonconventional treatments?

Of course. Dying patients are especially vulnerable to requests to try experimental and nonconventional treatments because of their own desire to prolong life, and because of the typical physician's view that death is a defeat. Professor Jay Katz has noted that in fighting against disease, "when medical knowledge and skills prove impotent against the claims of nature, physicians often resort to "disguised medical thinking." He continues, "At such times, all kinds of senseless interventions are tried in an unconscious effort to cure the incurable magically through a 'wonder drug,' a novel surgical procedure, or a penetrating psychological interpretation...doctors' heroic attempts to try anything may not necessarily be responsive to patients' needs but may turn out to be a projection of their own needs onto patients."[51]

Doctors often use the excuse that a dying patient has "nothing to lose" from taking part in an experiment. This is obviously not true from that patient's perspective, since the patient may have the dying process prolonged, experience more pain, or may even be rendered permanently incompetent or comatose. The most public experiments in the history of the world, those dealing with the artificial heart, have made it clear to everyone that there are fates worse than death, and that there are prices that are too high to pay even for added months of life.[52] Nor can experiments on dying children, such as transplanting a baboon heart into Baby Fae or doing multiple organ transplants, be justified solely on the basis that the child would die without the surgical intervention.[53] The scientific truth in each of these cases was that the child would die in any event, and the experimental intervention only dictated the manner of death.

The artificial heart experiments "have forced us to ask whether medical experimentation has become an acceptable form of euthanasia. Although society takes a negative view of a terminally ill patient who wants to end the suffering, it seems to be saying that it is acceptable for such patients to volunteer for experiments that could hasten their deaths."[54] The view these experiments on the dying expose is one that sees "death as abnormal and dying patients as subhuman" and casts "the terminally ill in modern rites of sacrifice" for the sake of science, putting the dying through torture in the vain hope of postponing the inevitable.[55]

The fact, of course, is that death is not optional; that both patients and physicians are vulnerable to "magical thinking" when the patient is threatened with death; and that only careful attention to both scientific facts and informed consent is likely to prevent serious exploitation of dying patients by ambitious researchers.

NOTES

1. Friesen & Kelly, *Do Cancer Patients Want to Be Told?* 27 Surgery 825 (1950); Samp & Currieri, *Questionnaire Survey on Public Cancer Education Obtained from Cancer Patients and Their Families,* 10 Cancer 382 (1957); Fitts & Ravdin, *What Philadelphia Physicians Tells Patients with Cancer,* 153 JAMA 901(1953); Oken, *What to Tell Cancer Patients?,* JAMA 1120 (1961); and Rennick, *What Should Physicians Tell Cancer Patients?,* 2 New Med. Materia 51 (1960).

2. *E.g.,* Novack *et al., Changes in Physician Attitudes Toward Telling the Cancer Patient,* 2, 41 JAMA 897 (1979). For an excellent discussion of telling the truth to terminally ill patients, *see* S. Bok *Lying* (New York: Pantheon, 1978), at 220–41; *and see* Kushner, "What Can I Tell You About Dying?" *Boston Globe Magazine,* Aug. 28, 1988, at 28.

3. *See* ch. XI, "Privacy and Confidentiality."

4. *See generally* ch. VI, "Informed Consent."

5. *See generally* ch. IX, "Human Experimentation and Research."

6. *In re Quackenbush,* 156 NJ Super. 282 (1978). For an excellent, up-to-date examination of the legal and ethical basis of the right to refuse treatment, *see* President's Commission for the Study of Ethical Problems in Medicine, *Deciding to Forego Life Sustaining Treatment* (Washington, DC: Government Printing Office, 1983).

7. *E.g., Lane v. Candura,* 6 Mass. App. 377, 376 N.E.2d 1232 (1978); *Bartling v. Superior Ct.,* 209 Cal. Rptr. 220, 163 Cal. App. 3d 186, (1984). And see cases cited at 318 New Eng. J. Med. 1755 (1988).

8. *See generally* Annas & Glantz, *The Right of Elderly Patients to Refuse Life-Sustaining Treatment,* 64 Milbank Mem. Q. 95 (Supp. 2, 1986), and cases cited therein; and *Deciding to Forego, supra* note 6.

9. *See generally* Annas & Densberger, *Competence to Refuse Medical Treatment: Autonomy vs. Paternalism,* 15 Toledo L. Rev. 561(1984).

10. *In re Yetter,* Northampton Co. Orphans Ct., No. 1973-533 (Pa. 1973) (Williams, J.).

11. *Lane v. Candura, supra* note 7.

12 *Id.* at 1235–36 *See also, Deciding to Forego, supra* note 6.

13. *See* discussion in ch. VI, "Informed Consent," at 89–91.

14. President's Commission for the Study of Ethical Problems in Medicine and Biomedical and Behavioral Research, *Making Health Care Decisions* (Washington, DC: Government Printing Office, 1982).

15. *Superintendent of Belchertown v. Saikewicz,* 373 Mass. 728, 370 N.E.2d 417 (Mass., 1977).

16. *See supra* notes 6 and 8, and Annas, *The Insane Root Takes Reason Prisoner,* 19 Hastings Center Report 29 (Jan. 1989).

17. G. J. Annas, L. H. Glantz & B. F. Katz, *The Rights of Doctors, Nurses and Allied Health Professionals* (Cambridge, Mass.: Ballinger, 1981), at 84.

18. The position of the American Medical Association, Council on Ethical and Judicial Affairs (Mar. 15, 1986) on "Withholding or Withdrawing Life-Prolonging Medical Treatment" is:

    The social commitment of the physician is to sustain life and relieve suffering. Where the performance of one duty conflicts with the other, the choice of the patient, or his family or legal representative if the patient is incompetent to act in his own behalf, should prevail. In the absence of the patient's choice or an authorized proxy, the physician must act in the best interest of the patient.

    For humane reasons, with informed consent, a physician may do what is medically necessary to alleviate severe pain, or cease or omit treatment to permit a terminally ill patient whose death is imminent to die. However, he should not intentionally cause death. In deciding whether the administration of potentially life-prolonging medical treatment is in the best interest of the patient who is incompetent to act in his own behalf, the physician should determine what the possibility is for extending life under humane and comfortable conditions and what are the prior expressed wishes of the patient and atttudes of the family or those who have responsibility for the custody of the patient.

    Even if death is not immanent but a patient's coma is beyond doubt irreversible and there are adequate safeguards to confirm the accuracy of the diagnosis and with the concurrence of those who have responsibility for the care of the patient, it is not unethical to discontinue all means of life-prolonging medical treatment.

    Life-prolonging medical treatment includes medication and artificially or technologically supplied respiration, nutrition or hydration. In treating a terminally ill or irreversibly comatose patient, the physician should determine whether the benefits of treatment outweigh its burdens. At all times, the dignity of the patient should be maintained.

    Approximately 80 percent of all physicians favor withdrawing life support systems from hopelessly ill or irreversibly comatose patients if they or their families request it (*American Medical News*, June 3, 1988, at 9). *See also* Hastings Center, *Guidelines on the Termination of Life-Sustaining Treatment and the Care of the Dying* (Bloomington: Indiana U. Press, 1988); and Glantz, *Withholding and Withdrawing Treatment*: The Role of the Criminal Law, 15 Law, Medicine & Health Care, 231 (1988).

19. *In re Osborne*, 294 A.2d 372, 375, n. 5 (D.C. Ct. App. 1972). *See also Bartling, supra* note 7, and the discussion of forcing treatment on pregnant women in ch. VIII, "Pregnancy and Birth, 126–30."

20. *In re Quinlan*, 70 N.J. 10, 355 A.2d 647 (1976).

21. *Saikewicz, supra* note 15. For a more detailed discussion of these two cases, *see* Annas, *Reconciling Quinlan and Saikewicz: Decisionmaking for the Terminally Ill Incompetent*, 4 Am. J. Law & Med. 367 (1979). These cases

have been followed on these points by other state supreme courts, and none has required routine court approval for such decisions.

22. *Cruzan v. Director, Missouri Dept. of Health,* 110 S.Ct. 2841 (1990), affirming *Cruzan v. Harmon,* 760 S.W.2d 408 (Mo. 1988) (en banc). *and see* Annas, *Nancy Cruzan and the Right to Die,* 323 New Eng. J. Med 670 (1990), and Rhoden, *Litigating Life and Death,* 102 Harv. L. Rev. 375 (1988).
23. Kutner, *Due Process of Euthanasia: The Living Will, a Proposal,* 44 Ind. L. Rev. 539 (1969). *See also* B. D. Colen, *The Essential Guide to the Living Will* (New York: Pharos Books, 1987).
24. *Deciding to Forego, supra* note 6. at 142.
25. *Id.* at 142.
26. *Id.*
27. *See* New York State Task Force on Life and the Law, *Life-Sustaining Treatment: Making Decisions and Appointing a Health Care Agent* (New York: Task Force, 1987).
28. *Id.,* and Charles P. Sabatino, *Health Care Powers of Attorney,* American Bar Association, Chicago Ill., 1990, (single copies free from AARP Fulfillment, [stock No. D138895], 1909 K Street, N.W., Washington, DC 20049).
29. *Prince v. Massachusetts,* 321 U.S. 158, 170 (1944).
30. Robertson, *Involuntary Euthanasia of Defective Newborns: A Legal Analysis,* 27 Stan. L. Rev. 213 (1975).
31. Robertson, *Dilemma at Danville,* 11 Hastings Center Report 5 (Nov. 1981).
32. *See* S. Elias & G. J. Annas, *Reproductive Genetics and the Law* (Chicago: Year Book, 1987), at 170.
33. *Bowen v. American Hospital Association,* 476 U.S. 610 (1986).
34. *Child Abuse and Neglect Prevention and Treatment Program: Final Rule,* 50 Federal Register 14877 (Apr. 15, 1985).
35. Elias & Annas, *supra* note 32, at 168 85.
36. *Maine Medical Center v. Houle,* Maine Superior Ct., Civil Action No. 74149 (1974); *and see* P. Stinson & R. Stinson, *The Long Dying of Baby Andrew* (Boston: Little Brown, 1983).
37. *Standards and Guidelines for Cardiopulmonary Resuscitation (CPR) and Emergency Cardiac Care (ECC),* 244 JAMA 453, 506 (1980).
38. *Matter of Dinnerstein,* 6 Mass. App. 466, 380 N.E.2d 134 (1978).
39. *Id.* For a case where the family's wishes regarding CPR overrode the directions of the patient against it, *see* E. Heron, *Intensive Care* (New York: Ivy Books, 1987), at 202
40. *See* Blackhall, *Must We Always Use CPR?,* 317 New Eng. J. Med. 1281 (1987); *and see* Ruark *et al., Initiating and Withdrawing Life Support: Principals and Practices in Adult Medicine,* 318 New Eng. J. Med. 25 (1988); Taffet, Teasdale & Luchi, *In-Hospital Cardiopulmonary Resuscitation,* 260

JAMA 2069 (1988); and Murphy, *Do-Not-Resuscitate Orders*, 260 JAMA 2098 (1988).

41. *Jane Hoyt v. St. Mary's Rehabilitation Center*, No. 774555, 4th Jud. Dist., Hennepin Co., Minn. aan. 2, 1981) (Arthur, J); *and see Judging Medicine, supra* note 22, at 212–16.
42. Blackhall, *supra* note 40.
43. Tomlinson & Brody, 318 New Eng. J. Med. 1758 (1988); *and see* Tomlinson & Brody, *Ethics and Communication in Do-Not-Resuscitate Orders*, 318 New Eng. J. Med. 43 (1988).
44. 1987 NY Laws 818; NY Public Health Law art. 29(B).
45. *See e.g.*, AMA and *Hastings Center Guidelines, supra* note 18; Levy, *Pain Control Research in the Terminally Ill*, 18 Omega 265 (1988).
46. Alsop, "The Right to Die with Dignity," *Good Housekeeping*, 1974, 69, at 130.
47 *See, e.g.*, D. Humphry & A. Wickett, *The Right to Die* (New York: Harper & Row, 1986), at 296–314. A proposed California initiative on this subject did not receive a sufficient number of signatures to get on the ballot in 1988. *See* Capron, *The Right to Die: Progress or Peril?*, 2 Euthanasia Rev. 41 (1987).
48. Louis Harris Poll, June 1987 (*Making Difficult Health Care Decisions, a survey conducted for the Loran Commission*); available from the Harvard Community Health Plan, Boston, Mass.
49. *See, e.g.*, Kuhse, *The Alleged Peril of Active Voluntary Euthanasia*, 2 Euthanasia Rev. 60 (1987), and other articles in this issue; Special Supplement, *Mercy, Murder, and Morality: Perspectives on Euthanasia*, 19 Hastings Center Report SS1-32 Oan. 1989), and Angell, *Euthanasia*, 319 New Eng. J. Med. 1348 (1988).
50. Schulman, *AIDS Discrimination: Its Nature, Meaning and Function*, 12 Nova L. Rev. 1113, 1140 (1988).
51. J. Katz, *The Silent World of Doctor and Patient*. (New Haven, Conn.: Yale U. Press, 1984), at 151.
52. *See, e.g.*, Annas, *Death and the Magic Machine: Informed Consent to the Artificial Heart*, 9 W. New Eng. L. Rev. 89 (1987).
53. *Judging Medicine, supra* note 22, at 384-90.
54. Brauer, "The Promise that Failed," New York Times Magazine, Aug. 28, 1988, at 76.
55. *Id.*, and *see* ch. IX, "Human Experimentation and Research."

XIII

Death, Organ Donation, and Autopsy

Death is the end of the journey for the patient, but many decisions must still be made by the patient's family and health care providers. If death occurs under circumstances that permit the deceased to be an organ donor, the family will likely be asked to consider donating one or more organs. Tissue donation, which does not require continuing circulation, will be an option in many more cases. Under some circumstances, an autopsy may be requested. In unusual circumstances, the medical examiner may want an autopsy performed to determine the cause of death. And ultimately, the family must be given the body so that they can cremate or bury it.

All states have now enacted a version of the "Uniform Anatomical Gift Act," and most amended their acts in the late 1980s. By following the provisions of this act, individuals can help their families make decisions about organ donation and autopsy by specifying what they would like done when they die. Everyone should make their wishes regarding organ donation known to their families.

When is a person "dead"?

Death has historically been a medical determination, and physicians have the legal authority to "declare" or "pronounce" a person dead. Traditionally, they have done so on the basis of irreversible cessation of respiration and heartbeat. It has only been since the early 1960s that a heart that had ceased beating could be restarted with cardiopulmonary resuscitation (CPR). But when CPR is either unsuccessful or untried, and the person's heart stops beating, the person is dead.

With the introduction of mechanical respirators (that breathe for a patient), it became possible to artificially sustain respiration and heartbeat in a body that would otherwise have ceased to function because the brain had been totally destroyed and thus could not even instruct the body to breathe. In order to prevent such dead bodies from wasting resources in intensive care units and to enable the use of the organs inside these bodies without committing homicide, a new way to determine death was proposed and has now been widely accepted: brain death.

What is brain death?

Brain death is a technical term that applies to bodies attached to mechanical respirators, whose brains have been totally and irreversibly destroyed, and who would therefore *never* be able to breathe on their own. The original definition was proposed in 1968 by the Harvard Ad Hoc Committee on the Definition of Death.[1] After more than a decade of public discussion and debate, the American Medical Association, the American Bar Association, the National Conference of Commissioners on Uniform State Laws, the President's Commission for the Study of Ethical Problems in Medicine and Biomedical and Behavioral Research, and many state legislatures endorsed the following language for the determination of death:

> An individual who has sustained either 1) irreversible cessation of circulatory and respiratory functions, or irreversible cessation of all functions of the entire brain, including the brain stem, is dead. 2) A determination of death must be made in accordance with accepted medical standards.[2]

It should be noted that part 2, the "brain death" part, is triply redundant to underscore the fact that the definition refers to the *entire* brain (including the brainstem), not just to the higher, or cognitive, functions of the brain. Thus, for example, individuals like Karen Ann Quinlan, who are permanently unconscious (called in a "persistent vegetative state") are not dead. This is because, among other things, they are capable of breathing without mechanical assistance, which demonstrates that their brainstem still functions.

This definition is biologically based (since the introduction of CPR, cessation of heartbeat has not necessarily meant irreversible destruction of the brain, but brain destruction always meant death) and widely accepted by the medical profession. Most states have laws that endorse it; nonetheless, "brain death" is a medical, not a legal, construct. Thus, physicians in *all* states may legally pronounce death on the basis of brain death criteria, so long as they do so in accordance with accepted medical standards.[3]

Does a patient's family have any say
in the determination of death?

No. No family member, judge, or legislature can bring the dead back to life. Determination of death is a medical decision to be made by application of accepted medical standards. If, however, the family has reason to doubt that a determination of death has been made in accordance with accepted medical standards, the family may properly insist that a qualified medical specialist (a neurosurgeon or neurologist) be called in to confirm the determination before the respirator is disconnected. Once the determination of death has been confirmed, however, the family has no right to insist that any "treatments" be continued.[4] All treatments should be ended upon the pronouncement of death (since a corpse cannot benefit from medical "treatment"), and the body released to the family for burial (unless organ donation or autopsy is planned).

Does brain death apply to children?

Yes. But it is more difficult to determine the irreversible absence of brain function in children, and special guidelines have been suggested for children in the following age categories: seven days to two months; two months to one year; and one to five years of age.[5] In children under one week of age, it is much more difficult to make an accurate determination of death based on brain criteria. This is one of the reasons that, logistically, it is extremely difficult and problematic to use breathing anencephalic newborns (babies born without higher brain function) as organ donors.[6]

How can an individual donate organs to others
so that they may have them upon the individual's death?

Under the Uniform Anatomical Gift Act (1987 revision), a version of which is law in every state, any person eighteen years or older and of sound mind may make a gift of all or any part(s) of his body to the following recipients for the following purposes:

1. Any hospital, physician, surgeon, or procurement organization for transplantation, therapy, medical or dental education, research, advancement of medical or dental science

2. Any accredited medical or dental school, college, or university for education, research, advancement of medical or dental science, or therapy
3. Any specified individual for therapy or transplantation or therapy needed by that individual

The gift can be made by provision in a will or by signing a card similar to the following in the presence of two witnesses:

UNIFORM DONOR CARD

Of _____

(name of donor)

In the hope that I may help others, I hereby make this anatomical gift, if medically acceptable, to take effect upon my death. The words and marks below indicate my desires. I give:

(a)_____ any needed organs or parts

(b)_____ only the following organs or parts

_____

(Specify the organ[s] or part[s])

for the purposes of transplantation, therapy, medical research or education;

(c)_____ my body for anatomical study if needed.

Limitations or special wishes, if any:

_____

This card is usually carried by the person signing it. In most states the gift can be revoked either by destroying the card or by an oral revocation in the presence of two witnesses.

Can an individual place conditions on the organ donation?

Yes. As the uniform donor card indicates, an individual can specify which organs are to be donated and the person or institution to whom they are to be donated. A donor may also specify how the body is to be buried following its medical use. The donee may accept or reject the gift. If accepted, and if only part of the body is donated, that part must be removed without unnecessary mutilation, and the remainder of the body must thereafter be turned over to the surviving spouse or other person responsible for burial.

Can a deceased's next of kin consent to donate organs if the deceased has not filled out a donor card?

Yes. Every statute so provides, and each state's statute also lists the order or priority of relatives who can consent in case close relatives are not available.

Everyone in the United States should be aware, however, that *even if a person signs a donor card, virtually no physician or hospital will take organs from the person's dead body without the consent of the next of kin.*[7] This is *not* because to do so would be illegal (it is not) or unethical. It is primarily because it does not seem "right" to most physicians. Hospitals must also deal with the living next of kin, and it is clumsy public relations to take organs from a deceased patient, whose body the next of kin has a right to bury, without getting the next of kin's agreement.

How many Americans have signed organ donor cards?

Despite intense public advertising for more than a decade, fewer than 20 percent of all Americans have signed organ donor cards.[8] One reason is that we simply do not like to think about our own deaths. But another is that Americans remain profoundly ambivalent about organ donation. Although 90 percent of the public say they "strongly approve of organ donation," for example, only about half that number have even discussed it with their families or would be willing to give permission to donate a family member's organ. Almost all neurosurgeons would donate a family member's organ, but only half of them have ever talked to their own families about organ donation.[9] There are also fears and misconceptions that inhibit signing organ donor cards. In a 1985 Gallup poll, for example, the following reasons were given for not signing such cards, in order of importance:

1. They might do something to me before I am really dead.
2. Doctors might hasten my death.
3. I don't like to think about dying.

4.   Family might object.
5.   Too complicated to give permission.[10]

What is "required request" and "routine inquiry"?

These are mechanisms designed to increase the number of organ donors in the United States. In the late 1980s the federal government and more than 40 states enacted "required request" legislation to require someone on the hospital staff to ask the next of kin of every potential organ or tissue donor whether or not they want to donate.[11] There are no penalties for noncompliance, and few of these laws are being followed.

An alternative suggestion is "routine inquiry" in which every patient is asked, upon admission to the hospital, if the person wants to be an organ donor. This was suggested as a standard amendment to the Uniform Anatomical Gift Act, but was defeated in 1987 because, among other reasons, it was thought that such routine inquiry might unduly upset hospital patients, especially those who had come for minor, elective procedures.

These suggestions seem well-meaning. But the law cannot force people to talk about a subject they do not want to talk about. Education concerning the need for organs, the use to which they will be put, and sensitive treatment of patients and their families is the most constructive way to increase organ donation.

Can a hospital refuse to release a body to the relatives if the relatives refuse to agree to organ donation?

No. Neither the patient nor the patient's relatives are under any obligation to donate organs. This is a voluntary act, and coercion has no place in organ procurement or "harvesting." Nonetheless, this type of extortion has been tried at least once.

A 1988 New Jersey case involved a twenty-year-old who committed suicide by shooting himself in the head.[12] He was rushed to the hospital and placed on a mechanical ventilator. Within an hour the emergency room physician informed his parents that their son was "brain dead" and asked them to consider donating his organs for transplant. Three hours later a neurologist confirmed death and again sought the parents' permission to use their son's organs for transplant. The parents were unable to decide and were asked to think it over and return in the morning with a decision. The next morning they informed the physician that they would not donate the organs, and that they wanted their son's body. The physicians refused to remove the mechanical ventilator and asked that they think more about organ donation. This charade went on for another two days before the hospital and physicians finally released the body to the parents, but only after they agreed to sign a hurriedly drafted and totally unnecessary release form that read:

> We have been advised by the attending physicians of our son, Jeffrey Strachan, that he has been declared "brain dead." It is therefore requested that all life support life-support-death [sic] devices be discontinued as soon as possible.
>
> In making this request we are fully aware of our legal responsibilities and further hold harmless John F. Kennedy Memorial Hospital and the attending physicians with regard to discontinuance of life support devices.

The New Jersey Supreme Court affirmed a jury decision that the hospital had wrongfully withheld the son's body, and that the parents had suffered emotional harm because the hospital had breached a duty it owed them as next of kin to release the body of their son for burial. In the court's words: "Although plaintiffs were told that their son was brain dead and nothing further could be done for him, for three days after requesting that their son be disconnected from the respirator plaintiffs continued to see him lying in bed, with tubes in his body, his eyes taped shut, and foam in his mouth. His body remained warm to the touch...a scene fraught with grief and heartache."

The actions of the doctors and hospital in this case were disgraceful and caused real suffering to real people. *Relatives should not be asked to consent to organ donation until after a determination of death has been made; if they refuse, mechanical ventilation must be immediately discontinued and the body released to them for burial.*

Must donors be screened before their organs or tissues are used in transplant?

Yes. Donors must be carefully screened to assure that their organs and tissues are suitable for transplant, and that they harbor no infectious agents, such as HIV. Both the procuring physicians and the hospital in which the transplant is performed may be held liable for damages if deficient screening results in injury to the recipient.

In one case, for example, two different patients received corneal transplants from the same donor. The transplanted corneas infected both eyes, resulting in total blindness. The patients sued. The eyes had been removed by a first-year ophthalmology resident, who had reviewed the incomplete chart of the deceased patient and had secured permission of the deceased's wife for the removal of his eyes. The hospital had no checklist that could be used as a guide, and the resident based his decision solely on what he had learned orally from senior residents.

At the trial, it was found that published criteria existed, which were fairly uniform around the country. These standards contraindicated use of any cadaver that had a history of certain medical conditions. The donor's record indicated he was a "60 year old white male, heavy alcoholic with cirrhosis of the liver proven at autopsy" and had suffered from several other serious diseases. The court concluded that "whoever may have had the responsibility of determining the suitability of the cornea for transplant would have been required, in the exercise of due care, to review carefully and exhaustively the medical history of the proposed donor." The jury could also have found the hospital itself was negligent "in failing to set up a procedure which would assure that the party responsible for determining the suitability of the cornea for transplant would have access to all of the relevant medical records of the proposed donor."[13]

Recipients of organ and tissue transplants have a right to rely on those who select the donor to properly screen for conditions that contraindicate use. This principle is applicable to all types of donations, from hearts and kidneys to sperm and blood.

Why are autopsies performed?

An autopsy is a comprehensive study of a dead body performed by a trained physician who employs recognized dissection procedures and techniques. The most common purpose of autopsies is to determine the cause of death, but they also serve valuable educational functions. Nevertheless, autopsy rates in United States hospitals have fallen from about 50 percent in the 1940s to 35 percent in 1972 to 10 to 15 percent in 1985. This decline is attributable to a variety of causes, including lack of reimbursement, better techniques to diagnose death, and the elimination of autopsy requirements by JCAH.[14] Usually only the thoracic, abdominal, and cranial cavities are opened during an autopsy, and neither the face nor the hands are cut or disfigured in any way.

Who is responsible for paying for an autopsy?

If a hospital doctor requests the autopsy from the family, the cost of the autopsy is almost universally absorbed by the hospital and does not appear as a separate item on the patient's bill.

Why might a family want an autopsy performed?

If the cause of death is potentially a genetic condition or infectious disease, exact determination may help other family members. Also, if the cause of death may have been related to medical malpractice, autopsy findings may help to prove this. The family may also simply wish to know the exact cause of death or may agree to an autopsy to contribute to medical education or research.

Who has the legal authority to consent to an autopsy?

Under the statutes of most states, the next of kin (the surviving spouse, if there is one, then other survivors in order of family relationship) has the right to consent to an autopsy. It has been ruled by courts on numerous occasions that there can be no property interest in a dead body. Nonetheless, the next of kin does have an interest in seeing that the deceased is properly buried, and this interest is strong enough to give the next of kin the right to possession of the body in the same condition it was in at death.[15] It is "the personal feelings of the survivors that are being protected" by the law.[16] Only an autopsy ordered by a state official can override this interest, and even in this case the decision must be made in good faith or the autopsy is illegal.

A hospital can be held liable in a civil law suit for refusing to deliver a body and, instead, inducing a coroner to perform an autopsy.[17] It can also be held liable, of course, for performing an autopsy without permission. In one case, the hospital was held liable for payment of damages for mental suffering when an autopsy was performed without consent, even though the widow did not detect the autopsy at the funeral. She found out about it only when she read the death certificate ten days later. The consent form had been signed by the doctor and two nurses as witnesses before being presented to the widow. She refused to sign it, but it was placed in the record anyway, and the doctor who performed the autopsy thought that permission had been granted because of the form.[18]

The extent of damages awarded for an unauthorized autopsy is not measured by the extent of the mutilation of the body but by the effect of the procedure on the "feelings and emotions of the [surviving relatives] who have the duty of burial."[19]

Can a patient consent to have an autopsy performed on his or her own body?

Laws in about half of the states specifically give the patient this right. Also, since under the provisions of the Uniform Anatomical Gift Act, a person may give all or a limited part of his body for education, research, and the advancement of medical science, the person may give his body to a hospital for the sole purpose of performing an autopsy. A person can make this consent binding on survivors by executing an instrument under the provisions of the Uniform Anatomical Gift Act, as explained elsewhere in this chapter. The probable reason doctors seldom ask patients to consent in advance to autopsies is that it seems callous. Moreover, few hospitals would permit the autopsy, even with this authorization, over family objection.

Can the hospital retain any portion of the body after autopsy?

No. The hospital cannot retain any tissues or organs without the permission of the person who consented to the autopsy.[20] Standard forms for permission generally include permission to retain samples, and the rule against retention does not cover limited small portions of organs taken for further study. The person consenting to the autopsy, however, has the right to place whatever limitations on the consent he or she wishes.

The reasonable expectation of the public is that larger portions of the body will not be retained, even for research, without consent.

For example, over a three-year period an associate medical examiner in Milwaukee removed and stored the testicles from about seventy males on whom she had performed autopsies. The incident received much publicity. She defended the practice in the name of medical research (although she had not commenced any on the testicles she had collected), saying, "In view of the fact that testicles are removed during an autopsy, I don't see why it is a crime, as has been implied, to put them in a jar instead of back in the body."[21] The answer, of course, is that it is the individual and the next of kin who have the right to decide how the body will be disposed of, not the medical examiner.

A hospital could also be liable if it misplaces an entire body. In a Florida case, a jury awarded $150,000 in damages to a couple when a hospital lost the corpse of their premature baby. The mother testified, "I still have my doubts that Paul ever died."[22] This case was, however, overturned on appeal. The court ruled that recovery could be had only on proof of "wantonness, wilfulness or malice" on the part of the hospital.[23]

Do health care professionals or students have a right to practice their skills on a brain dead corpse?

No. This may be done only with the prior written consent of the patient or with the consent of the next of kin given *after* the patient has been declared dead. Before death, only the patient has the right to consent to nonbeneficial research or training on his body; after death the relatives, who have the right to bury the body and care for it before burial, have this exclusive right. Any unauthorized use or mutilation of the body after death is illegal and unethical. Fresh corpses are frequently used to practice such things as vaginal exams and intubations.[24] Even though it is often better for patients that dead bodies rather than live ones are used for such practice, consent must be obtained in each instance.

NOTES

1.   Ad Hoc Committee of the Harvard Medical School to Examine the Definition of Brain Death, *A Definition of Irreversible Coma*, 20 JAMA 337 (1968).

2. President's Commission for the Study of Ethical Problems in Medicine and Biomedical and Behavioral Research, *Defining Death* (Washington, DC: Government Printing Office, 1981).
3. *See* G.J. Annas *Judging Medicine* (Clifton, NJ.: Humana Press, 1988), at 365–69. Those who believe that judges or legislatures can postpone or defer death by fiat are simply incorrect as a matter of both law and fact.
4. New York State health regulations seem to provide the patient or family some say in choosing the basis on which death will be determined. This provision is an irrational political compromise that can only breed confusion (NY Public Health Law art. 29(B) [1987]).
5. Task Force for the Determination of Brain Death in Children, *Guidelines for the Determination of Brain Death in Children*, 21 Annals of Neurology 616 (1987)
6. *See* Annas, *From Canada with Love: Anenephalic Newborns as Organ Donors?* 17 Hastings Center Report 36 (Dec. 1987); and Capron, *Anencephalic Donors: Separate the Dead from the Dying*, 17 Hastings Center Report 5 (Feb. 1987).
7. Task Force on Organ Transplantation, *Organ Transplantation* (Washington, DC: HHS, 1986). The selling of organs has also been outlawed in the United States. *See Judging Medicine, supra* note 3, at 378-83.
8. *Id., Hearings on Organ Transplants: Hearings before the Subcomm. on Investigations and Oversight of the House Committee on Science and Technology*, 98th Cong., 1st Sess. (April 13, 14, 27, 1983).
9. Prottas & Batten, *Health Professionals and Hospital Administrators in Organ Procurement: Attitudes, Reservations and their Resolutions*, 78 Am. J. Public Health 642 (1 988).
10. *Organ Transplantation, supra* note 7, at 38.
11. Annas, *Paradoxes of Organ Transplantation*, 78 Am. J. Public Health 621 (1988).
12. *Strachan v. John F. Kennedy Memorial Hospital*, 109 NJ 523, 538 A.2d 346 (1988).
13. *Ravenis v. Detroit General Hospital*, 63 Mich. App. 79, 84, 234 N.W.2d 411, 414 (1975).
14. Landefeld, Chren, Myers, *et al., Diagnostic Yield of the Autopsy in a University Hospital and a Community Hospital*, 318 New Eng. J. Med. 1249 (1988).
15. *Infield v. Cope*, 58 N.M. 308, 270 P.2d 716 (1954); *Gahn v. Leary*, 318 Mass. 425, 61 N.E.2d 844 (1945).
16. *Strachan, supra* note 12.
17. *E.g., Darcy Presbyterian Hospital*, 202 NY.259, 95 N.E. 695 (1911).
18. *French v. Ochsner Clinic*, 200 So. 2d 371 (La. App. 1967).
19. *Id.* at 373, and Strachan, *supra* note 12; and *see* Annas, *The Cases of the Live Buried Mother and the Dead Unburied Baby: Negligence or Outrageous Conduct?*, 5(3) Orthopaedic Rev. 71 (Mar. 1976).

20. For a discussion of the legal and ethical issues involved in the commercial use of human tissues and cells, *see* US Congress, *Office of Technology Assessment, New Developments in Biotechnology: Ownership of Human Tissues and Cells* (OTA-BA-337) (Washington, DC: Government Printing Office, 1987); and Annas, *Whose Waste Is It Anyway? The Case of John Moore,* 18 Hastings Center Report 37 (Nov. 1988).
21. Rosenberg, "Examiner Says She Will Stop Stealing Testicles from Dead," *Milwaukee Journal,* June 26, 1979, at 1. And *see* "Doctor Admits That He Sold Corpses' Parts," *New York Times,* Sept. 4, 1988, at 50.
22. Boston Globe, Nov. 28, 1973, at 2.
23. *Brooks v. South Broward Hospital District,* 325 So. 2d 479 (Fla. Dist. Ct. App. 1975), and *see supra* note 19. A related issue is the use of tissue from dead fetuses following elective abortion. The laws of most states permit the use of fetal tissue for research and transplantation, but only with the consent of the mother. Sale of fetal tissue is prohibited by federal law. *See generally* Robertson, *Fetal Tissue Transplants,* 66 Wash. U.L.Q. 443 (1988).
24. Orlowski, Kanoti & Mehlman, *The Ethics of Using Newly Dead Patients for Teaching and Practicing Intubation Techniques,* 319 New Eng. J. Med. 439 (1988) (this article mistakenly advocates use of dead bodies "in the absence of expressed dissent").

XIV
Medical Malpractice

The major trend in medicine in the past two decades has been toward the transformation of the physician's primary goal from treating patients in the best way they know how to the goal of treating patients in a way that minimizes their potential exposure to a medical malpractice suit. This risk-aversive movement is characterized by use of "defensive medicine" (tests and procedures ordered not to help the patient but to protect the physician in the event of a lawsuit), almost constant complaining about the price of liability insurance and the "medical malpractice crisis," and the rise of "risk management" as a health care industry specialty. Many physicians feel that the medical profession has lost control over its own destiny, and that regulators, lawyers, and insurance companies have stripped them of their professional autonomy. Physicians have a point: Public policy over the past decade has been driven by the desire to contain and reduce costs rather than the desire to increase access to medical care or to improve its quality. This emphasis has tended to limit physician income and hurt physician morale. There is no magic solution to the "medical malpractice insurance problem." This chapter describes the current system, outlines some of its costs and benefits, and suggests that patients and their physicians can work together to establish a partnership that will help physicians (in cooperation with their patients) regain professional autonomy and will simultaneously decrease the likelihood of lawsuits being filed for unanticipated negative results of treatment.

What is medical malpractice?

Medical malpractice denotes the basis for a lawsuit by a patient against a health care provider for injuries suffered as a result of the provider's negligence. The method for compensating victims of medical malpractice is a fault-and-liability system through which the person at fault is responsible to pay for the harm inflicted on an innocent victim. Whether the health care provider is at fault is determined in an adversary proceeding in which the provider and the patient are each

239

represented by legal counsel. The trier of fact, usually a jury, must decide whether the health care provider is responsible for the injury.

Primarily because they are usually decided by lay juries in public, physicians have deplored malpractice lawsuits for almost a century and a half. In 1845, for example, physicians indicated alarm at the increase in malpractice lawsuits and suggested alternatives to jury trials, such as committees made up of physicians, to judge such claims.[1] In 1872, the American Medical Association recommended that physicians be appointed independent arbiters by the court to judge their peers.[2] Physicians have also historically hated the term "malpractice" itself, a term that denotes "evil" or "bad" practice. In fact, it refers simply to a physician who has not lived up to the customary professional standard set by the actions of the "average competent physician" in the same or similar circumstance. Today, more than 80 percent of all malpractice suits are brought on the basis of an incident that occurred in a hospital. About 15 percent involve doctors' offices, and the rest occur in nursing homes, HMOs, surgical centers, and other settings. Our society permits malpractice suits for three basic reasons: (1) to control quality by holding health care providers accountable for their actions, (2) to compensate patients for injury, and (3) to give patients an opportunity to express dissatisfaction with the care they have received.

What elements must a patient prove to win a malpractice claim?

A valid malpractice claim against a health care provider must have four elements: duty, breach, causation, and damages. Each element must be proven by a "preponderance of the evidence" (that it is more likely than not that it is true). *Duty* to a patient requires the prior establishment of a provider–patient relationship and is defined by the standard of care. The standard by which a provider's actions are measured is that of a reasonably prudent practitioner under the same or similar circumstances.[3] *Breach* of that duty by specific conduct on the part of the practitioner, by action or inaction, is measured by the applicable standard of care. *Proximate cause* denotes a causal connection between provider's conduct and the damages alleged by the patient, that is, the provider's breach of duty must be the cause of the patient's harm. The plaintiff cannot recover any money damages for improper conduct on the part of the defendant if the breach of duty itself produced no harm or injury. The final element is the actual injury or *damages* suffered by the patient, and these are measured in monetary terms.[4]

Do health professionals guarantee patients a favorable outcome of treatment?

Generally not. Usually the contract between the provider and the patient does not include a guarantee of a good result.[5] But it is possible for a provider to change the terms of the contract and specifically include such a guarantee.

For example, in one case the doctor treated a patient for ear trouble, telling him that he should undergo stapes mobilization operations. The doctor allegedly advised the patient that, even though his condition might not be improved by the operations, his hearing would not be worsened as a result. After three operations, the patient's hearing was much worse. The patient based his lawsuit on the breach of an express contract. The trial court concluded that the facts alleged were sufficient to find that there was an enforceable contract, and the Supreme Court of Kansas, on appeal, agreed.[6]

Can patients hold all physicians to the same standard of care regardless of where they practice?

Traditionally, the "locality rule" was applied so that the standard of care required of a nonspecialist physician was that commonly possessed and practiced by other physicians in the same community or similar communities.[7] Some courts even restricted the locality rule to the same community. The reason given for these restrictions was that there were significant differences between the facilities and opportunities for consultation in different places, particularly between large cities and those in rural areas.[8]

With progress in medical education, exchange of medical knowledge, and availability of medical centers, and ability to refer patients to them, the rationale for the locality rule no longer exists. Through the years, the rule has been altered; in some states, it has been modified, and in others, it has been completely abandoned.[9] This is the modern trend. In most jurisdictions, the local standard of practice may be considered by the jury as a factor, but will not be the sole determinant. Expert witnesses can usually be used from any part of the United States to testify on the standard of care.

What is the standard of care
to which a patient can hold a specialist?

The law imposes a higher duty on a specialist than on the general practitioner. By definition, a specialist is a physician who devotes special attention to a particular organ or area of the body, to the treatment of a particular disease, or to a particular category of patients. The specialist is required to possess that degree of knowledge and ability and to exercise that amount of care and skill ordinarily possessed and exercised by physicians practicing in the same specialty, regardless of geographic location.[10] This rule is substantially identical in all states and applies to all specialties.[11]

How can a patient establish at trial that a provider failed
to meet the recognized standard of medical practice?

Ordinarily, for the patient to prevail, he or she must present the testimony of *an expert medical witness* (a licensed physician) to establish that the provider failed to fulfill his duty, and that as a result, the patient was injured. This is because the standard of care is that of the reasonably prudent provider, and in most cases only another provider has sufficient "expert" knowledge to establish that standard. The expert witness explains what the health care community recognizes as the standard of care in a particular situation and further gives an opinion whether or not the provider–defendant's conduct met that standard of care. A jury of lay people is not competent to know the standard of care in a health matter, or whether the defendant complied with it. Therefore, as a matter of law the plaintiff's case will be dismissed unless the necessary expert testimony is presented.[12]

When the case involves a specialist, usually the expert witness is from the same specialty. As long as the subject matter of the case is common to both individuals, a specialist in one area may be permitted to testify against a specialist in another.[13] Similarly, a specialist may testify to the standard of care of a general practitioner as long as the specialist is familiar with the applicable standard of care.[14]

What is *"res ipsa loquitur"*?

The general rule that requires expert testimony in malpractice cases does not apply in situations where the negligence is so obvious that lay people can determine it without the help of expert witnesses. This rule, called *res ipsa loquitur,* "the thing speaks for itself," is applicable when all three of the following conditions are met:

1. An injury has occurred of a type that does not ordinarily occur in the absence of negligence.
2. The instrumentality or conduct that caused the injury was, at the time of the injury, in the exclusive control of the defendant.
3. The plaintiff was not guilty of contributory negligence.[15]

The *res ipsa loquitur* doctrine has been mainly used in cases involving foreign objects left in the patient after surgery, burns from heating equipment, and injury to a part of the patient's body outside the treated area.[16] For example, the Minnesota Supreme Court ruled that *res ipsa loquitur* applied when a surgeon, who was in exclusive control of a scalpel that broke during an operation, left part of it in the patient.[17] A number of states have passed statutes modifying the doctrine in medical malpractice actions. Most of these statutes restrict its application to enumerated situations, such as foreign objects, explosions, and burns.

When may a patient's malpractice suit be barred by the statute of limitations?

A statute of limitations sets a time period within which a person must commence a particular type of lawsuit. This is to ensure that adjunction takes place soon after the occurrence of the alleged negligence, so that witnesses and evidence are more likely to be available, and that people will not have to go through life with the threat of lawsuits being brought against them for things done long ago.

Many states have statutes of limitations specifically applicable to medical malpractice actions. If not, the statute applicable to tort actions will usually govern. A lawsuit not started within the specified time period is barred and will be dismissed on a motion by the defendant. State laws differ on the length of time provided for starting a lawsuit and on when the time period begins to run, but most have a one- to three-year statute of limitations on medical malpractice suits, and two years is common.

The time period begins to run when the elements of a lawsuit occur: when the negligent act allegedly occurs; when the doctor–patient relationship or continuous series of treatments ends; or when the harm to the patient is, or reasonably should have been, discovered (the *discovery rule*).

The modern trend is to apply the discovery rule. It is a rule of common sense: How could the injured patient be expected to bring a lawsuit when the patient neither knew nor reasonably could have known about the injury? Some states limit application of the discovery rule to cases involving allegations of foreign objects left in patients following surgery. In other states, it has been held inapplicable in cases involving misdiagnosis,[18] treatment,[19] and the incorrect administration of a blood test.[20] Some states, however, extend the rule to all cases of medical malpractice.[21]

A statute of limitations may be "tolled," or prevented from running, for a variety of reasons. The most common ones are that the injured patient is a minor,[22] is insane,[23] or is in the armed services; that the defendant is absent from the state[24]; or that the defendant has fraudulently concealed the basis for the suit.[25]

Almost all states have taken legislative action to redefine their statutes of limitations to restrict the number of lawsuits. Statutes of limitations specifically applicable to malpractice suits have been shortened; statutes have been enacted in states where none existed; and the application of the discovery rule has been curtailed. The trend is to legislate shortened statutory time periods for medical malpractice actions.

What is a contingency fee?

Most lawyers handle medical malpractice cases on a contingency fee. Under this fee system, a lawyer is paid only for out-of-pocket expenses unless the suit is won, in which case the lawyer takes about 25 to 50 percent of the award as payment for services. This payment system is often blamed for contributing to the number of malpractice claims. It is argued that contingency fees encourage lawyers to pursue claims of doubtful merit or to require unjustifiably large amounts in terms of potential damages for legitimate claims, in hope of achieving recovery through settlement or awards from sympathetic juries.

Legislation has been enacted in many states to regulate plaintiff-attorney fees. These have taken various forms. One requires that the court review an attorney's proposed fees and approve what it considers "reasonable fees." Several set a fixed percentage ceiling for contingency fees in malpractice actions. Others adopt a sliding scale, most often expressed in terms of a percentage of the final award. Under this arrangement, as the amount recovered increases, the lawyer's percentage decreases.

In attempting to establish reasonable guidelines for the amount that a plaintiff's attorney can receive from the injured patient's award, these statutory provisions perform a needed service. But to the extent that they reduce the number of claims brought by diminishing the willingness of attorneys to handle certain meritorious claims, they are a disservice to those injured patients who cannot otherwise afford legal counsel.

The contingency fee structure also compels lawyers to screen out claims that are spurious, or for which recovery appears less than probable, and to refuse claims for which damages would not amount to enough to cover their expenses. Since the attorney, rather than the plaintiff, bears the financial risk of losing the suit, the attorney has no incentive to invest any time or money in a claim for which recovery appears doubtful.[26] In addition, with the average unregulated fee rate approximately one-third of recovery, many lawyers decline malpractice cases that will probably achieve settlements or awards of less than $25,000, because the expected compensation for the amount of time expanded is not seen as worthwhile. The threshold value for the acceptance of cases for which recovery is less than probable would, on the average, be higher, perhaps as much as $250,000.

What is a pretrial screening panel?

A pretrial screening panel is an informal procedure designed to screen out nonmeritorious claims quickly and inexpensively and to encourage the settlement of other cases amicably. More than thirty legislatures have established review plans by statute. Additionally, there are a small number of voluntary, nonstatutory mediation plans sponsored by state or local medical societies and bar associations.

The typical pretrial review applies to any malpractice action brought against a health care provider, regardless of the size of the claimed damages. The panel consists of three to seven members, including at least one attorney, one health care provider, and frequently one consumer member. The hearing itself is almost always informal. The parties have the choice of accepting the panel's decision, negotiating their own settlement, or rejecting the decision of the panel and proceeding to court.

When court action is undertaken, and the panel's decision is in favor of the patient, the panel is often obligated to help the patient obtain the necessary expert medical testimony for trial. On the other hand, the party that loses the panel decision may face various "penalty" provisions for deciding to disregard the panel decision and go to court. For example, some states provide that the findings of the panel are admissible at a later trial of the case. Some statutes require the party rejecting the panel decision to post a cost bond with the court, which is payable if the opposing party prevails at trial.

Can a health care provider require a patient to sign a binding arbitration agreement before treatment?

Generally not. Arbitration is a nongovernmental procedure for settlement of disputes between private parties. Parties to a dispute submit their differences to the judgment of an impartial person or panel appointed either by mutual consent or by statute or court decision. Imposed arbitration is generally applied to disputes under a certain maximum amount, with the jury trial system preserved for the larger cases.

More than a dozen states have passed legislation specifically providing for binding arbitration of medical malpractice claims by written agreement of the parties, and malpractice claims can be arbitrated in at least thirty states under the general arbitration statute. Arbitration is sometimes offered in group health care organizations, such as HMOs and other prepaid plans, where the agreement to submit malpractice claims to binding arbitration is part of the subscriber's contract. Because an individual is not required to become a member of such a group plan, the acceptance of the obligation to arbitrate is considered voluntary.

What are the problems with the current medical malpractice system?

One's view of the problems of the present system depends almost exclusively on one's perspective. The current system has been described as in "crisis;" a crisis primarily related to insurance availability and cost.

Insurance companies have argued that the precipitous rate increases of the mid-1980s were required by increases in the frequency and severity of malpractice litigation. They have cited figures indicating sharp increases in the absolute numbers of malpractice suits filed and in the number of million-dollar-plus verdicts returned by juries. The companies have charged that America has become an increasingly lawsuit-prone society, and that premium increases are needed to keep pace with increases in litigation expenses and unrealistic jury verdicts. As they did in the mid-1970s, the insurance companies in the mid-1980s again successfully sought changes in the legal system to make it harder for injured patients to seek compensation from physicians and hospitals.[27] And even though there was no more substance to their argument than there was in the 1970s, almost all states enacted new medical malpractice and tort reform acts in 1986 and 1987. A few states, including Florida and New York, also increased the authority of the insurance commissioner to freeze or regulate malpractice insurance rates. By mid-1988, the second "crisis" was over,[28] but debate on changing the current tort system continues.

The American Medical Association (AMA) has tended to side with the insurance companies and focus its efforts on "reforming the tort system." The AMA's four major recommendations for change are (1) limiting jury awards for noneconomic damages ("pain and suffering") to $100,000; (2) requiring deductions in awards for money that victims get from any other source (eliminating the collateral source rule); (3) using mandatory periodic payments for all awards for future damages that exceed $100,000; and (4) limiting attorney fees by modifying the contingency fee to a sliding scale.

Lawyers have argued that such changes are misplaced. The American Bar Association (ABA), for example, rejects the AMA's recommendations, arguing that the real problem with medical malpractice *is* medical malpractice. Accordingly, the ABA recommends that physicians set up much tougher methods of policing themselves to eliminate incompetent and impaired physicians and thereby protect the public. The ABA notes that changes in the legal system were tried in the 1970s and failed as a method to deal with the problem, and that this demonstrates that the real problem is not with the legal system. If the problem is the way we handle personal injury suits in this country, then the ABA argues we should change the tort system across the board and not just make adjustments for physicians.

As for the AMA's specific proposals, trial lawyers generally argue that limiting "pain and suffering" awards to $100,000 is unfair to severely injured victims for whom this amount cannot begin to compensate; and requiring reduction of awards for outside insurance benefits penalizes those who carry such insurance and is a windfall to the wrongdoer. Finally, they note that radically changing the contingency fee will close the courtroom to the poor and middle class, who cannot afford to pay a lawyer out of their own pocket to bring a complex and costly lawsuit.

Consumer groups, like Ralph Nader's Public Citizen, continue to view the professional debate between physicians and lawyers as a sideshow. They note that it is the insurance companies that have dramatically raised the rates for insurance coverage, and that these companies continue to exploit the fear of physicians to reap profits. Stock market values of the insurance companies, they have noted, more than doubled in 1985, the same year in which these companies drastically increased their premiums. In fact, although property/casualty companies suffered a $46 billion underwriting loss from 1975 to 1984, they also had about $121 billion in investment gains during this period, for a net profit of about $75 billion.

This is probably why many think the insurers are too comfortable and need to be much more closely monitored and regulated. The president of the National Insurance Consumer Organization, Robert Hunter, for example, has argued that insurance premiums are much more closely linked to the industry's profit cycle (which is determined primarily by their return on investment) than to legal liability. He noted that in 1975, at the beginning of the 1970s' insurance crisis, the industry earned only 4 percent on its reserves. Premiums skyrocketed in 1976 and 1977 to make up for this loss. The premium increase brought sharp profit increases (similar to those now being experienced), which attracted a flood of new capital and led to premium-price cutting to accumulate more cash to invest at high interest rates. When these rates fell, the industry earned only 2 percent on equity in 1984. As a consequence, and a repeat of 1976–77, premium rates shot up for 1985–86. The only difference was that other insurance lines were affected as well, including liability insurance for day care centers, cities, and many products. It is completely implausible, Hunter notes, to argue that the legal system, as represented by judges and juries, worked perfectly well from 1977 to 1983 and then all of a sudden went out of control in 1984 and 1985.[29]

The consumer groups seem correct. Although the number of million-dollar-plus verdicts increased to over four hundred in *all* personal injury cases in 1985, these were verdicts, not payments or settlements, which are usually significantly lower. Other studies have concluded that the total number of civil lawsuits is *not* increasing per capita in the United States, and that the median (middle value) jury award has actually remained constant for the past twenty five years, at about $20,000.

Economists have found the situation mixed. There is general agreement that the system is inefficient at compensating victims of malpractice, but is cost-effective as a method of quality control.[30] Its inefficiency at compensation stems from two facts: Very few people who are actually injured by malpractice ever get into the system; and there are very high transaction costs that go to the lawyers and insurance companies (who get more of the total premiums than injured victims do). Studies suggest that approximately 1 of every 100 patients admitted to a hospital suffers an injury because of medical negligence. Nonetheless, only between 1 in 25 and 1 in 50 of these injured patients is ever compensated through the current system. But even though compensation is very limited, as long as the potential for a suit deters any meaningful percentage of negligent injuries, the system will be cost-effective from the patient's viewpoint.

From a systemwide perspective, malpractice litigation is actually inexpensive. Malpractice premiums for physicians *and* hospitals combined amount to less than 1 percent of total expenditures on health care. And although some specialists pay higher rates, the average physician, even in 1988, still spent less than 5 percent of gross income on malpractice insurance, a percentage that has not changed significantly during the past decade. Of course, there is plenty of room to make the system more cost-effective and more efficient, and some specialists, such as obstetricians, are being charged unrealistically high insurance premiums in some states.[31]

What changes in the current medical malpractice system make sense from the patients' point of view?

Some people have suggested looking abroad for new ideas. Sweden and New Zealand are seen as lands of no lawsuits, and some have urged that we adopt their "no fault" systems, which compensate everyone for injury, regardless of cause, by paying their medical bills. What this solution misses, of course, is the fact that these countries, like every industrialized country in the world, except the United States and South Africa, have a system of national health insurance in which everyone's medical bills are paid, regardless of cause or source of injury or illness. Likewise, Sweden has a vast social support system that is government-financed.

In early 1986, the General Accounting Office issued a report to Congress entitled *Medical Malpractice*. Its subtitle, "No Agreement on the Problems or Solutions" still describes the situation. In April 1986, the *New York Times,* in a lead editorial, depicted the crisis in insurance availability (all insurance, not just medical malpractice) as one with four possible solutions: "(A) crack down on the lawyers who manipulate juries to win outlandish settlements and fat contingency fees; (B) crack down on the insurance industry, which seems unable to manage its cash flow responsibly; (C) try to teach the public that enormous liability judgments are not cost-free; (D) all of the above."

The *Times* thinks the correct answer is "D" and chided the Reagan administration for concentrating only on "A" in its proposal to reform products liability tort law. In medical malpractice as well, the answer is "all of the above." Specifically, the legal profession must discipline its members who file frivolous lawsuits and pursue untenable claims, and it should do so vigorously. States and the federal government should regulate insurance companies, so that accurate data on premiums and claims paid can be compiled. Reasonable premium raises should be announced well in advance, limited on a yearly basis, and cancellations permitted only with adequate notice. In addition, we need more effective methods to prevent incompetent and impaired physicians from hurting patients and more effective communication between physicians and their patients.

How is medical malpractice litigation related to informed consent?

Although physician concern for lawsuits is not new, the public's perception of what is possible in medicine, and medicine's power over disease, has changed dramatically and in some cases naively. This perception is shared by both patients and their physicians and has been spawned by the arrival of the "new technological age" of medicine. Technology has given us, as novelist Don Delillo puts it, "an appetite for immortality."[32] We believe that diseases *can* be controlled, and that physicians *should* be able to do something for us when we fall ill. We want to believe that we can have it all; live our lives without regard to physical or mental dangers and then go to the "repair shop" (the hospital) when we suffer a physical or mental "breakdown" and have it fixed. We have adopted the image of ourselves that commentators on industrialization have feared for decades: We see ourselves as machines, and physicians as mechanics. If physicians cannot repair us, it must be because they lack the skill, do not know the latest techniques, or make a mistake.

The major problem with medical malpractice insurance is medical malpractice insurance. But unrealistic expectations on the part of patients, and ritualistic silence and demands for blind faith on the part of physicians, exacerbate the broader medical malpractice situation greatly. *Enhancing the doctor–patient partnership by taking informed consent and shared decision making seriously could go a long way toward "solving" the medical malpractice problem.* This will require that uncertainty in medical diagnosis and treatment be acknowledged by both patients and physicians. This acknowledgment alone should radically decrease the felt need for "defensive medicine." Other solutions deserve fair hearings, but should be judged against the three primary goals of the current tort system: compensation for injury, quality control, and responsiveness to consumers. Only changes that enhance one or more of the goals, without offsetting losses in others, deserve serious consideration.

How can state licensing boards be strengthened to help protect patients from incompetent and impaired physicians and nurses?

All physicians and nurses must obtain a state license before they can "practice medicine" or "practice nursing." The general requirements for such a license are the possession of a specified academic degree (such as an MD or OD), successful completion of a written examination, and "good moral character." Once the licensing agency (usually known as the "Board of Registration in Medicine" for physicians and the "Board of Registration in Nursing" for nurses) grants the license, it almost never takes it away and usually has no formal method for monitoring actual practice. The inability of these state agencies to protect the public from impaired professionals has been a source of much public criticism. This should not be surprising in an era when even professional associations agree that 5 to 10 percent of all health care providers are impaired by drugs, alcohol, or psychological problems. The public has historically (and accurately) viewed these licensing agencies, which are dominated by members of the profession rather than members of the public, as existing primarily to protect doctors and nurses, rather than to protect the public.

It is unusual for a patient to even think about complaining to a state licensing agency about an allegedly incompetent or impaired physician or nurse. This is unfortunate, because although one must prove four elements to win a malpractice suit (duty, breach, damages, and causation), the licensing board need only prove the first two (duty and breach, that is, that the individual failed to exercise the same reasonable prudence that a qualified professional would have in the same or similar circumstances) to take action against a professional.

State licensing boards could be greatly improved if the public was given all or at least a majority of the positions on the board; if they were adequately staffed to deal with complaints; if they effectively informed the public how to bring complaints; and if they demonstrated to the public that complaints would be handled fairly and expeditiously, and that the process and results would be made public. In addition, licensing boards that do not already have it should be given the authority to order psychiatric examinations on allegedly impaired physicians and drug- and alcohol-screening tests on physicians for whom there is reason to believe that their addictive behavior is endangering patients. These tests should be followed up by mandatory treatment programs and limited, supervised practice for a period of time after treatment, if the practitioner desires to keep his license to practice.

What penalties can a licensing board impose?

A licensing board can censure a practitioner, suspend a practitioner's license for a period of time, or revoke a license to practice altogether. Some licensing boards also have the authority to impose fines on a practitioner. Boards can also act creatively by conditioning continued licensure on fulfilling requirements to do specific things, such as retraining in a specialty, continuing to obtain psychiatric or substance-abuse help, or doing community service.

Since licenses are granted for life, theoretically a professional need not open a book or medical journal after licensure. Thus, a physician who graduated from medical school in 1950 may still be practicing and using 1950 knowledge, which is the equivalent today of what using leeches would have been in 1950. To help prevent this, most licensing boards require physicians and nurses to take "continuing education" courses to keep up on new developments. But since no tests are given, there is no way to evaluate the effectiveness of continuing education. It would be much more protective of the public if physicians and nurses were required to take a formal examination to retain their license, at least every ten years. A proposal for relicensure by examination has been made in New York, and although it will be extremely controversial among licensed professionals (who will not want to be retested because they distrust the ability of the test to

measure their own ability, and because they may not pass), it merits
strong public support.[33]

How can a patient file a complaint against a physician or nurse?

Citizens can file a complaint in writing with the state Board of
Medicine (for physicians), Board of Nursing (for nurses), or Board
of Dentistry (for dentists). The names and addresses of these agencies can be obtained from the office of the state attorney general.[34]
Although patients will not receive any money should the complaint
be found meritorious and the health care provider disciplined,
patients are helping to prevent the health care provider from hurting
other patients. This is an important role of citizens, and *if you believe your physician or nurse might injure others, you should file a
complaint with the licensing board.* It should set forth the facts as
you know them, together with all documentation you have. In addition, you should be prepared to release relevant medical records to
the board and to discuss your case with the board or representatives
of the board. You may also file a complaint against a hospital with
the agency that licenses hospitals, usually the Department of Public
Health. You need a lawyer to file a lawsuit. *You do not need a lawyer to file a complaint with a licensing agency.*

How does a person find a lawyer to bring a malpractice suit?

It is not easy. Finding a good lawyer is at least as hard as finding
a good doctor. When looking for a lawyer who specializes in medical malpractice and personal injury law, perhaps the two most important qualities are experience and compassion. Ask friends about lawyers they have worked with, and if you have a family lawyer, ask
this person as well. You can also get names from the local bar association, the local office of the American Civil Liberties Union
(ACLU), and from some of the organizations listed in appendix A of
this book. It is important to feel comfortable dealing with your lawyer,
because medical injury is such a personal and traumatic experience,
and the lawsuit itself may take years to resolve. You should interview
at least three lawyers before deciding on which one is right for you.
Experience is extremely important, but as important is compassion
and confidence on your part that the lawyer will treat your case as a

priority and be available to discuss it with you as it moves along. Remember, while you will be expected to pay expenses, you do *not* pay for the lawyer's time. The lawyer gets paid only if the case is won or settled and then will get a percentage of the settlement. This percentage (usually 25 to 50 percent) should be made explicit at the time you retain the lawyer and put in writing.

What is the future of the tort-liability system as a means for resolving medical malpractice claims under national health insurance?

In general, full health care coverage under one of the proposed "comprehensive benefits" national health insurance plans should reduce the number of malpractice suits, although not necessarily with a corresponding reduction in the occurrence of medical negligence. Those persons who suffer injuries allegedly because of negligence, and who primarily require additional medical care, will probably not sue their health care provider, since the cost of the original and any additional medical care will be covered. But eliminating a large number of private malpractice suits, and therefore much of the "policing" function that these actions perform, will put additional strain on either internal or administrative regulatory procedures to prevent the occurrence of medical negligence. The tort-liability system affects the quality of medical care, the compensation provided to injured patients, and the responsiveness of the health care system to the consumer. To the extent that national health insurance will limit recourse to this system, additional procedures, such as the patient rights advocate discussed in the following chapter, must be designed and implemented to respond to these issues. Medical malpractice litigation will likely retain an important role in any future health care system in the United States.[35]

NOTES

1. Burns, *Malpractice Suits in American Medicine before the Civil War*, 43 Bull. Hist. Med. 41, 52 (1969). But *see* Bennet, "Pluses of Malpractice Suits," New York Times Magazine, July 24, 1988, at 31.
2. D. Konold, *A History of American Medical Ethics: 1847–1912* (1962), at 50 51. For contemporary perspectives *see Health Care Improvement and*

Medical Liability (Proceedings of an HHS Research Conference) (Washington, D.C.: HHS, 1988) and A. Holder, *Medical Malpractice Law*, 2d ed. (New York: John Wiley & Sons, 1978).

3. In extremely rare cases, where the entire medical profession or speciality has failed to keep up with medical advances, the courts themselves will define "reasonable prudence." *See, e.g., Helling v. Carey*, 519 P.2d 981 (Wash. 1974) (failure to do routine glaucoma test is negligent as a matter of law).

4. *E.g., Lab v. Hall* 200 So. 2d 556 (Dist. Ct. App. Fla. 1967).

5. *E.g., Hill v. Boughton*, 146 Fla. 505, 1 So. 2d 610 (1942).

6. *Noel v. Proud*, 189 Kan. 6, 367 P.2d 61 (Kan. 1961).

7. *E.g., Williams v. Chamberlain*, 316 S.W.2d 505 (Mo. 1958).

8. *Michael v. Roberts*, 91 NH 499, a3 A.2d 361 (1941).

9. *E.g., Brune v. Belinkoff*, 354 Mass. 10 a, 235 N.E.2d 793 (1968); *Murphy v. little*, 112 Ga. App. 517, 145 S.E.2d 760 (1965); *Douglas v. Bussabarger*, 73 Wash. 2d 476, 438 P.2d 829 (1968); *Blair v. Eblen*, ad 370 (Ky. 1970). *See also Tallbull v. Whitney*, 172 Mont. 6, 564 P. 2d 167 (1977).

10. *E.g., Francisco v. Parchment Med. Clinic*, 407 Mich. 325, 285 N.W.2d 39 (1979) *Robbins v. Footer*, 553 F.2d 1 a3 (D.C. Cir. 1977).

11. *E.g., Barnes v. Bovenmyer*, 255 Iowa 220, a N.W.2d 312 (1963); *Lewis v. Read*, 80 NJ Super. 148, 193 A.2d 255 (1963); *Belk v. Schweizer*, 286 N.C. 50, 149 S.E.2d 565 (1966); *Siirila v. Barrios*, Mich. App. 72, 2a8 N.W.2d 801(1975), *aff'd*, 398 Mich. 576, 248 N.W.2d 171 (1976).

12. *Sims v. Helms*, 345 So. 2d 7 al (Fla. 1977); *see also Buckroyd v. Bunten*, 237 N.W.2d 808 (Iowa 1976); Marsha Tomaselli, 118 R. I. 190, 372 A.2k 1280 (1977).

13. *Radman v. Harold*, 279 Md. 167, 367 A.2d 472 (1977). But *see Callahan v. William Beaumont Hosp.*, 400 Mich. 177, 254 N.W. 2d 31 (1977).

14. *Siirila v. Barrios*, 398 Mich. 576, 248 N.W.2d 171 (Mich. 1976).

15. *E.g., Mondot v. Vallejo Gen. Hosp.*, 15 App. 2d 588, 313 P.2d 78 (1957); *Irick v. Andrew*, 545 S.W.2d 557 (Tex. Ct. Civil App. 1976).

16. *E.g., Seneris v. Haas*, 45 Cal.2d 811, 291 P.2d 915 (1955).

17. *Young v. Caspers*, 311 Minn. 391, 249 N.W.2d 713 (1977).

18. *Robinson v. Weaver*, 550 d 18 (l'ex. 1977).

19. *Proewig v. Zaino*, 394 NYS.2d 446, 57 A.D. 892 (1977).

20. *Simmons v. Riverside Methodist Hosp.*, 44 Ohio App. 2d 146, 336 N.E.2d 460 (1975).

21. *Sanchez v. South Hoover Hosp.*, 132 Cal. Rptr. 657, 553 P.2d 1129 (1976); Moran v. Napolitano, 71 N.J. 133, 363 A.2d 346 (1976).

22. *Graham v. Sisco*, 248 Ark. 6, 449 ad 949 (1970); *Chaffin v. NiCosia*, 621 Ind. 698, 310 N.E.2d 867 (1974).

23. *Miller v. Dickert*, 190 S.E.2d 459 (S.C. 1972).

24. *Swope v. Printz*, 259 S.C. 1, 468 ad 34 (1971).

25. *Nardone v. Reynolds,* 538 F.2d 1131 (Sth Cir. 1976).
26. Annas, Katz & Trakimas, *Medical Malpractice Litigation under National Health Insurance: Essential or Expendable?,* 1975 Duke L. J. 1335, 1344; *and see* Dietz, Baird & Beru, *The Medical Malpractice Legal System,* in US Dept. of Health, Education, and Welfare, *Report of the Secretary's Commission on Medical Malpractice,* App., at 87, 119 (1973).
27. The best book on the crisis of the mid-1970s is S. Law & S. Polan, *Pain and Profit: The Politics of Malpractice* (New York: Harper & Row, 1978). On the 1980s, *see* Danzon, *The Effects of Tort Reforms on the Frequency and Severity of Medical Malpractice Claims,* Ohio St. L. J. 413 (1987); Note, *The Constitutionality of Medical Malpractice Legislative Reform: A National Survey,* 18 Loy. U. Chi. L. J. 1053 (1987); and Note, *1986 Tort Reform Legislation: A Systematic Evaluation,* 73 Cornell L. Rev. 628 (1988).
28. "Insurance Crisis Is Over—At Least for Now," *Hospitals,* Apr. 20, 1988, at 46. *See also* Harrington & Litan, *Cause of the Liability Insurance Crisis,* 239 Science 737 (1988); and Blum, "Malpractice Claims Down, Costs Are Up," National Law Journal, Jan. 23, 1989, at 1.
29. New York Times, Apr. 13, 1986, III, at 3.
30. P. Danzon, *Medical Malpractice* (Cambridge, Mass.: Harvard U. Press, 1985)
31. *See, e.g.,* D. K. Roberts, J. A. Shane & M. L. Roberts, eds., *Confronting the Malpractice Crisis: Guidelines for the Obstetrician-Gynecologist* (Kansas City, Mo.: Eagle Press, 1985); and K. S. Fineberg *et al., Obstetrics/Gynecology and the Law* (Ann Arbor, Mich.: Health Administration Press, 1984).
32. D. Delillo, *White Noise* (New York: Penguin, 1986), at 285.
33. *See also* Note, *The 1985 Medical Malpractice Reform Act: The New York State Legislature Responds to the Medical Malpractice Crisis,* 52 Brooklyn L. Rev. 135 (1986).
34. A complete list of the names, addresses, and phone numbers of these agencies appears in C. B. Inlander, L. S. Levin & E. Weiner, *Medicine on Trial* (New York: Prentice-Hall, 1988), at 239–53. Under the Health Care Quality Improvement Act of 1986, and the Medicare and Medicaid Patient and Program Protection Act of 1987, the federal government will operate a national data bank to keep track of physicians who have been disciplined and move from one state to another. Patients will *not* have access to this data, although they should. *See* Gianelli, "Data Bank to Chronicle Licensing, Malpractice Actions,"*American Medical News,* Jan. 13, 1989, at 11.
35. For an excellent discussion of medical malpractice litigation in Great Britian, *see* Miller, *Medical Malpractice Litigation: Do the British have a Better Remedy?* 11 Am. J. Law & Med. 433 (1986). *See also* Annas et al., a6; and Abraham, Medical Liability Reform: A Conceptual Framework, 260 JAMA 68 (1988).

XV
The Patient Rights Advocate

Recognition of patient rights in the form of a Patient Bill of Rights is a necessary, but not sufficient, step in the protection and promotion of patient rights. Rights are not self-actualizing. We may look forward to the day when all physicians, nurses, and allied health professionals will accord patients their basic human rights as a matter of course, but this day has not yet come. Until this goal is attained, mechanisms that can help to ensure that patient rights are protected and honored will be absolute necessities. Some of these mechanisms are discussed in this chapter. I continue to favor the one Jay Healey and I proposed in 1974: the patient rights advocate.[1] This and other approaches to protecting the rights of patients are also discussed here.

The notion of patient rights was novel in the early 1970s. Historians have, however, already put it in context. Paul Starr, for example, discusses the patient rights movement as part of the "generalization of rights" in the United States. In addition to the movement (still unfulfilled) to recognize health care as a basic human right was the more attainable movement to work for rights *in* health care, "such as the right to informed consent, the right to refuse treatment, the right to see one's own medical records, the right to participate in therapeutic decisions...For every right there are always correlative obligations... Recognized rights in health care, such as informed consent, obligate doctors and hospitals to share more information and authority with their patients. Thus, *the new health rights movement went beyond traditional demands for more medical care and challenged the distribution of power and expertise*" (emphasis added).[2]

As Starr goes on to note, increasing skepticism with the assumption that health providers always know best and always act out of concern for their patients, led to the emergence of "a variety of legal safeguards aimed at limiting professional autonomy and power" and an attempt to "demedicalize critical life events, such as childbirth and dying."[3] Courts contributed mightily to this trend, as is demonstrated throughout this book. But patients should not have to go to court to have their rights vindicated. The doctor–patient partner-ship should be a standard way of relating in the doctor–patient rela-tionship. When it is not, the patient needs an advocate to maintain human rights and dignity. Just as the 1960s and 1970s were a time of defining patient rights, the 1980s and 1990s are a time to make these rights a reality in medical care.

What is a patient rights advocate?

A patient rights advocate is a person whose job is to help patients exercise the rights outlined in the state's or institution's Patient Bill of Rights. The advocate may be employed by the health care facil-ity, prepaid health plan, an insurance company, a government agency, a consumer group, or the patient. The critical characteristic is loyalty: *The patient rights advocate must represent the patient.* This is essential because the goal is to enhance the patient's posi-tion in making decisions, not to encourage the patient to follow facility routine or to "behave."

More than three thousand hospitals now employ individuals with a job title of "patient representative." This title can be misleading, since their real job is often not to represent the patient but, rather, to represent the institution that employs them. I have been harsh on these individuals in the past, probably too harsh, calling them "hospital representatives." There certainly are some who do an excellent job of helping patients in spite of their potential conflict-of-interest posi-tion. My current view is that the patient should give the hospital's "patient representative" a fair chance to help; but as soon as it becomes clear that the patient representative is unable to help or is more con-cerned with protecting the hospital than with protecting the patient, other avenues of redress must be explored.

The problem of misleading labels is well illustrated in Studs Terkel's classic book, *Working*. One of the workers he interviewed was Betsy Delacy, a "patients' representative" in a 540-bed hospital. Ms. Delacy described her job as admitting the patient to the hospital, following the patient throughout the stay, and afterward making sure the bill is paid. In her words: "I don't feel I represent the patient. I represent the hospital. I represent the cashiers. I'm the buffer between the patient and the collection department...We visit our patients as

often as we can, so they get to know us as their representative. 'Are
you comfortable?' 'Are you satisfied with your food?' Then, when
he gets to know me—'I know your account is going to be a problem...'
I'm not looking for money, but if the patient doesn't ask such
questions, I mention it. I sort of joke with 'em and then lay it out and
sock it to'em." [4]

It seems reasonable to ask why such a person is called a "patients'
representative" in the first place. Ms. Delacy explains, "It seems
strange that you should have a collection department in a hospital.
Patient representative has a better sound. Nobody knows what it's
all about. It's like any organized business. They give people such
titles that nobody knows what it's all about." Unfortunately, Ms.
Delacy seems to be correct. Few people seem to know what being a
real patient representative or patient advocate is all about. This is
unacceptable. Hospitals are not like "any organized business," and
patients are not like any other consumer—they are sick and extremely
vulnerable. They need advocates who can help them with their medically related problems, not an expert in housekeeping functions (that
should be supplied, along with clean linen, as a matter of course).
Without scrupulous attention to their rights as human beings, their
personhood will be stripped from them, and they will be treated as if
they were pets.

Patient representatives may also be recruited from the public relations department, and if their duties are limited to nonmedical issues,
such a "representative" is a pure public relations gimmick, a "happy
hostess." [5] Patients need real advocates, and if the hospital does not
supply one, patients will have to bring their own: a friend, lawyer,
physician, nurse, social worker, or relative. In theory, all of these
people are qualified to perform the job, because all of them can help
the patient exercise the rights specified in the Patient Bill of Rights.
On the other hand, a formal advocate system in the hospital is potentially much more powerful because the advocate can have direct access to other members of the staff, administration, and relevant committees in the hospital structure, and can develop credibility in problem solving with the staff. No matter who a patient decides can properly act as his or her advocate, it is critical that each patient have one.

What powers should the patient rights advocate have?

A patient rights advocate is an individual whose primary responsibility is to assist the patient in learning about, protecting, and asserting his or her rights within the health care context. It is essential that the specific rights of patients be *spelled out in a bill of rights that the hospital adopts as policy and which the advocate has the power to enforce.* The word advocate is used in its classical sense, *advocare,* "to summon to one's assistance, to defend, to call to one's aid." Connotations of adversariness, contentiousness, and deliberate antagonism are both unfortunate and unnecessary. *The goals of a patient rights advocate system are to:*

1. Protect patients, especially those at a disadvantage within the health care context (for example, the young, the severely handicapped, those with AIDS, the poor, the incommunicative, those without relatives, those unable to speak English).
2. Make available to patients the opportunity to participate actively with the doctor as a partner in a personal health care program.
3. Restore medical technology and pharmaceutical advances to their proper perspective by confronting the exaggerated expectations of the modern American medical consumer.
4. Reflect in the doctor–patient relationship the reality of the health–sickness continuum and assert the humanness of death as a natural and inevitable reality.

The advocate should be able to exercise, at the direction of the patient, rights and powers that belong to the patient. They include:

1. *Complete access to medical records* and the authority to make notes in the medical record regarding patient complaints and demands
2. *Active participation in hospital committees* responsible for monitoring the quality of care within the health care context, especially utilization review, patient care, risk management, and quality assurance
3. *Access to support services for all patients who request them*

4. *Participation at the patient's request and direction in discussion of the patient's case*
5. *Ability to delay discharges*
6. *Ability to call in medical consultants to aid or advise the patient*
7. *Ability to consult with the hospital's legal counsel and to lodge complaints directly to hospital director and executive committee*

It is also very useful for an advocate who works for the hospital to have the authority to write off disputed hospital bills. An example of how an advocate with these powers can help patients occurs regularly in hospitalization for childbirth. Under intense pressure from insurance companies, the "standard" length of stay for maternity is continuing to shrink in the United States, down to twenty-four to forty-eight hours in many hospitals for an "uncomplicated delivery." Longer stays can be arranged, but may require a statement from a physician and contacting the insurance company directly. A woman who had just given birth and was suffering from postpartum bleeding was severely pressured by the nursing staff to leave. They "hassled" her, continually asking her when she was going, saying the utilization review was "bothering them." Even though she felt very sick, they told her she was "ready to leave" and was just "a little nervous." With no one to turn to, she felt obliged to leave the hospital and go home. In her words: "Because I was so exhausted, I just could not advocate for myself. So I went home. I had no help. I could barely walk. My husband got sick immediately, and it took me literally months to recover [from childbirth]."[6]

There is no excuse for this type of shoddy treatment on either legal or medical grounds. The existence of an effective advocacy system in which the advocate could review medical records, call in a consultant, and delay discharge could put a quick stop to it. An advocate should be available to patients 24 hours a day, 7 days a week, since problems do not occur only from nine to five during the day.

Most of the criticism directed against the patient rights advocate has involved the alleged introduction of conflict into the hospital setting. This seems less a reaction to the concept itself than a reaction to one way of carrying out the responsibilities. The empowered patient and the advocate both confront the hospital and resist the exercise of arbitrary authority. The relegation of all serious decision making to adversary proceedings would raise serious questions. But this is neither the goal nor the likely outcome of the advocate model. The goals are set forth above, and even when adversariness does develop, the advocate will improve the doctor–patient relationship by promoting openness and honesty and identifying real problems so they can be dealt with effectively.

What is usually meant by the term "ombudsman" in the hospital setting?

The ombudsman is an alternative approach, although both an advocate program and an ombudsman could function in the same facility. The ombudsman's role is to seek out broad problem areas, to research facts, to publicize grievances to appropriate audiences, and to make suggestions about resolving those problems. An ombudsman does not participate in the actual resolution. The result is active representation in problem identification without direct personal influence on its solution. Such an approach eliminates the potential problems created by an adversary system. The danger, however, is that the ombudsman would have no influence on important decisions.

Another suggestion is to combine the best aspects of both approaches while discarding those aspects that would have a detrimental effect on its proper functioning. Part of the patient rights advocate's function could, for example, be as an ombudsman with respect to protecting the rights of patients. While remaining available to respond to all patients who desired services, the advocate could also review incident reports, compile lists of recurring situations in which the patient rights are affected, classify them according to seriousness, and take action by publication and by making suggestions for changes when warranted.

What qualifications should an advocate have?

The advocate must have a basic knowledge of medicine—know the language and how to read medical records—and law. The advocate must also be able to communicate with patients, nurses, doctors, the hospital lawyer, and the hospital administration. Persons chosen for this role should be overqualified rather than underqualified in terms of both knowledge and community acceptance. More than half of the current patient representatives have a bachelor's or master's degree, and almost two-thirds have been previously employed in the same hospital as a nurse, social worker, admissions department worker, or other category.[7] There are no data available to judge which background produces the most effective advocate for patient rights. There must also be a sufficient number of advocates to ensure patient access to their services.

Who should have the power to hire and fire the advocate?

Financing and supervising an advocate system is a complex problem. Like most money problems, however, it is primarily a question of priorities. The rights of patients have not been treated as high priority in many health care facilities. The only way to ensure that advocates are not transformed from patient to management representatives or risk managers is to have them hired by someone outside the facility (or possibly by the board of trustees) and be primarily accountable to the patients they serve. In the future, an advocate program might be built into HMO and other managed care contracts (in which case the advocate could work for the consumer-dominated board of directors), be required as a condition of JCAH accreditation or participation in Medicare (or any forthcoming program of national health insurance), or be funded by a statewide consumer agency. For now, health care facilities are likely to be the primary employer. As long as advocates see the job as requiring that their primary loyalty be to the patient, they can be helpful to patients no matter who pays them. Indeed, in any "enlightened" health care facility, the administration will recognize that serving patients and respecting their rights are primary goals, and that everyone is better served and more legally secure in an environment that prizes and promotes patient rights.

How would a patient rights advocate improve health care while safeguarding human rights?

Examples of how such a system would improve patient care and enhance human rights are legion. When he presented his patient rights advocate proposal to the trustees of Boston City Hospital in the early 1970s, Dr. David F. Allen, chief resident in psychiatry and president of the House Officers Association, told them of the time he was called to the emergency ward to talk with a woman whose stomach had just been pumped out. He was the first person on the scene able to speak Spanish to the Puerto Rican woman. She told him that she had had some very distressing news at home, had taken two Alka Seltzers, and had come to the hospital to talk with someone. The staff at the emergency room had assumed she was an overdose case "because most Puerto Ricans who demonstrate symptoms like those shown by the woman have overdosed."[8] An advocate with the capability of speaking Spanish could have prevented this routine treatment.

Senator Edward Kennedy recounts the following incident.

> Paul, a ten year old boy, had a seizure at his home and passed out. His father picked him up and rushed him to a police station. The nearest hospital was a private institution. Paul had been receiving treatment at the County Hospital which was some distance away. The police said they could not take him there because it was out of their district. When they arrived at the hospital, Paul's father was subjected to an interview about his finances and insurance. No one would look at Paul until his father had answered such questions as: "Do you own your own home?" "Who do you work for?" "How long have you worked there?" The interviewer also refused to call the County Hospital. In frustration, Paul's father left the emergency room at the hospital and drove the long distance to the County Hospital. In the course of his trip, he passed several hospitals but was afraid to stop because of the possibility that they would treat him as the first hospital had. He arrived at the County Hospital where his son died within an hour.[9]

The case illustrates the tragic results that occur when a hospital places housekeeping chores above medical duty in an emergency situation. An advocate could have promptly asserted the legal right of the patient to receive immediate emergency care without reference to ability to pay. Failure to provide an opportunity for the right to be asserted was a significant factor in the loss of a life. An advocate could have played a key role in saving it.

In another case, a college professor wrote of his own experiences. He had been admitted to a hospital for diagnosis:

> I got a reinforcement of the sense of not only am I a patient who is supposed to behave in a certain way, but I'm almost an object to demonstrate to people that I'm not really people any more, I'm something else. I'm a body that has some very interesting characteristics about it....I began to feel not only the fear of this unknown, dread thing that I have, that nobody knows anything about—and if they know, they're not going to tell me—but an anger and a resentment of "Goddamn it, I'm a human being and I want to be treated like one!" And feeling that if I expressed anger, I could be retaliated against, because I'm in a very vulnerable position.[10]

Some of his frustrations would find an outlet in the person of the patient rights advocate. The advocate would be a person whom the patient could talk to without fear of retaliation; a person who could pull his medical records and tell him whether or not a diagnosis had been made; a person who could take the time to explain the diagnostic tests or find someone who could; a person who could explain why medical students were present, and that the patient could have them excluded if he or she wished; a person who could assure the patient that no matter what the patient's attitudes toward the medical staff or expressions of fear and resentment were, no retaliatory action would be taken against him or her. Tension and conflict would be reduced and the quality of medical care improved.

Joan Haggerty's experiences with childbirth illustrate more routine problems:

> She and her husband had attended classes on natural child birth. They had discussed the matter with the doctor in the out-patient clinic of the hospital where the child would be delivered. The hospital had a policy of allowing the husband in the delivery room "at the doctor's discretion." They entered the hospital and spent three hours together in the labor room. As she was being transferred to the delivery room the doctor (a resident) said to the husband, "Sorry, you can't come in, you make me nervous."
>
> In the delivery room Ms. Haggerty, who had previously given birth by the natural method in England, demanded that the stirrups

be removed. The attendants laughed at her and held her down as her wrists were strapped to the table by leather thongs.[11]

Under a patient advocate system, with an advocate assigned to the maternity ward, the advocate would be in charge of advising the medical personnel about the couple's desires concerning natural childbirth, would make whatever preparations were deemed necessary, and be present at the parents' request to be sure that the father was not denied access to the delivery room during birth (for example, by assuring an alternative obstetrician is available). The advocate would also see to it that the mother was not subjected to coercion or ridicule during birth. (This latter function would probably not be necessary if the husband were allowed to be present in the delivery room as a matter of course.) Doctor–patient relationships in all of these examples would be improved by the advocate.

What is a "risk management" program?

Most hospitals have set up "risk management" programs to help control and limit their exposure to financial losses, especially through medical malpractice claims. These programs, if well run, help educate all hospital personnel to issues of patient rights and patient safety, and benefit patients generally.[12] Many of these managers, however, view the job of protecting the institution as somehow separate and distinct from the job of protecting patients. And when risk managers withhold records from patients, or information about injuries caused by the hospital staff is concealed, they are engaged in active deception of patients and deserve nothing but contempt.

Responsible risk managers, on the other hand, can help patients by informing them of any medical mishap that may have injured them or compromised their care, and by taking steps to minimize its long-term impact on the patient's health. They also serve a valuable educational function when they help to raise the awareness level of patient rights in the hospital. Like patient representatives, risk managers should initially be given the benefit of the doubt. If the patient or the patient's family thinks a potentially dangerous or harmful situation exists, and that harm can be avoided by taking action, the risk manager may be able to help—such as by calling in consultants if no one else will. Also, after an injury has occurred, the risk manager may be a useful source of information, at least if the risk manager understands that protecting individual patients and honoring their rights are the most effective and constructive ways to also protect the hospital's integrity and reputation.

What are "ethics committees" and "ethics consultants"?

Many large hospitals have either or both of these mechanisms. The ethics committee is a group of physicians and others (often including the chaplain, a lawyer, a nurse, and a social worker). Ethics committees help develop hospital policy in "ethical" areas (many of which are actually legal) such as "do not resuscitate" policies and policies on withholding or withdrawing treatment from handicapped newborns and other incompetent patients; consult on specific cases; and run educational programs on ethics in the hospital. A few such committees also make recommendations on how individual cases should be resolved.[13] A functioning advocate system could make some of these committees obsolete and unnecessary. Nonetheless, if the patient or the patient's family is having difficulty implementing a patient care decision because the doctors or nurses think the decision is "wrong" or "unethical" or "illegal," it can sometimes be very useful to ask that the case be reviewed by the hospital's ethics committee. If the committee agrees with the patient, the patient's wishes will likely be carried out. If the committee disagrees, the patient is no worse off than when the exercise began, and the advocate can still resort to other mechanisms, including the courts, if necessary.

An ethics consultant is a person, instead of a committee. The consultant is usually trained in philosophy, religon, or medical ethics but could be a physician with an interest in ethics. If the hospital has such a person, the ethics consultant can often be very helpful in resolving conflicts between patients and medical staff. Patients and their families should not hesitate to call on the ethics consultant for help when conflicts cannot be resolved by simply getting everyone involved in the care of the patient to sit down and talk to each other. It should be stressed, however, that perhaps 90 percent of all treatment conflicts are resolvable (and often resolved) in a "team meeting" in which all the caregivers (including physicians and nurses) and all family members (including the patient, if competent) meet to discuss the facts of the case and present their reasons for their differing recommendations on treatment. Such a conference should be tried first.

What is the "It's Your Fault" ploy?

Martha Lear chronicled her physician–husband Harold Lear's experience with four heart attacks and more than a dozen hospitalizations in her book *Heartsounds*. In the hospital, depersonalization was vividly portrayed and devastatingly destructive. Residents were routinely uncaring and authoritarian. One refused Lear's request for a milder pain medication, saying, "If you want what I ordered, you can have it. If not, you'll get nothing." Another resident gave him a huge dose of potassium before he had any food or water; the incident produced a stomach ulcer. And whenever he got worse or had a problem, health care professionals blamed him for it. The Lears called this response the "It's Your Fault" ploy:

> Why did the operation take so long?
> Because you lost so much blood.
> *Not:* Because the surgeon blew it.
>
> Why do you keep making these tests?
> Because you have a very stubborn infection.
> *Not*: Because I can't diagnosis your case.
>
> Why did I get sick again?
> Because you were very weak.
> *Not:* Because I did not treat you competently the first time.[14]

Dr. Lear constantly asks himself if he treated patients this way, and usually admits that he did. He suggests that every physician be required to spend at least a week a year in the hospital bed: "That would change some things in a hurry." And most observers agree that although physicians make terrible patients, physicians who have been patients are almost always better physicians because of the experience. When Lear views the hospital from a patient's perspective, he sees it as a vast slaughterhouse:

> He thought with awe of the tension between this hushed hospital atmosphere and the things that were happening here: in this polite, muted, ordered way, people were being torn open, ripped

apart, bones cracked, holes made and tubes stuck into holes, and they moved passive as cattle along the conveyor belt—oh, yes, passive, not even breathing for themselves; the tube that had been in his throat was like the slaughterhouse hook, embedded in the animal's throat, by which it, he, was hung.[15]

The "It's Your Fault" ploy is played out constantly in hospitals, and patients and their advocates need to be alert to it. It is not meant callously; it is almost done by rote, as if blaming the victim is actually part of the medical procedure. We all know the response to the question, "Why are you having so much trouble drawing blood?" "Because your veins are so small. (*Not*: Because I'm not very good at it).

What steps should health care facilities take now to enhance patient rights?

All health care facilities should adopt a simple five-point agenda that would greatly enhance patient rights and help humanize the hospital environment. These five points are:

1. Eliminate "routine" procedures.
2. Provide patients open access to their medical records.
3. Provide for twenty-four-hour-a-day visitation.
4. Require full experience disclosure before procedures are performed.
5. Implement an effective patient rights advocate program, which includes a patient-centered Bill of Rights.[16]

The patient rights advocate has already been discussed. The other points merit brief explanation:

1. *No routine procedures.* It is common for nurses and others to respond to the question "Why is this being done?" with "Don't worry, it's routine." This is not an acceptable response. Procedures should not be performed on patients simply because they are routine; they should be performed only because they are *specifically* indicated for a patient. Thus, routine admission tests, routine use of johnnies, routine use of wheelchairs for in-hospital transportation, routine use of sleeping pills, to name a few notable examples, should be abolished.

Use of these procedures treats patients like fungible robots rather than persons; and "routine" procedures are often demeaning and unnecessary.

2. *Open access to medical records.*[17] Although currently required by federal law and many state statutes and regulations, open access to medical records by patients remains difficult, and a patient often asserts the right to see the record at the peril of being labeled "distrustful" or a "troublemaker." The information in the hospital chart is about the patient and properly belongs to the patient. The patient must have access to it, both to enhance his own decision-making ability and to make it clear that the health care facility is an "open" institution that is not trying to hide things from the patient. Surely, if facility personnel make decisions about the patient on the basis of information in the chart, the patient also deserves access to the information, even in the absence of a specific law or health care facility policy on this subject.

3. *Twenty-four-hour-a-day visitor rights.* One of the most important ways to humanize the health care facilities and enhance patient autonomy is to assure the patient that at least one person of the patient's choice has unlimited access to the patient's room at any time of the day or night. This person should also be permitted to stay with the patient during any procedure (for example, during childbirth or induction of anesthesia), so long as the person does not interfere with the medical care of other patients.

4. *Full experience disclosure.* The most important gain of the past decade has been the almost universal acknowledgment of the need for the patient's informed consent.[18] Nevertheless, an important piece of information that is material to the patient's decision is still routinely withheld: the experience of the person doing the procedure. Patients have a right to know whether the person asking permission to draw blood, do blood gases, do a bone marrow aspiration or a spinal tap (to list just a few examples) has ever performed the procedure before, and if so, what the person's rate of adverse effects is. This applies not only to medical students and student nurses but also to board certified surgeons. We all do things for the first time, but not every patient wants to take such an active role in education.

We have begun the long journey toward humanizing health care facilities and promoting patient self-determination. But more specific measures are needed before patients will be assured that they can effectively exercise their rights in institutional settings.

What steps can individuals take to promote patient rights?

Perhaps the most important step is to educate yourself about your rights. The second is to exercise your rights regularly so the health care professionals get used to the idea that citizens in the United States take their rights seriously and are not willing to abdicate them simply because they are sick. Third, you can support legislation that codifies and strengthens patient rights, making them easier to understand and easier to exercise.

Two model bills are suggested in the appendixes of this book: A Patient Bill of Rights Act (appendix B) and a Right to Refuse Treatment Act (appendix D). Passage of these acts in each state would help make sure that basic human rights are afforded to the sick and hospitalized. Appendix A provides a list of additional resources for those interested in more specific information about the topics covered in this book. Appendix E provides a brief introduction to the medical and legal literature for those who want to do their own research.

Patient rights are important to everyone. Insist that they are respected and take steps to ensure that the physicians you trust with your body treat you as a full partner in treatment decisions.

NOTES

1. Annas & Healey, *The Patient Rights Advocate: Redefining the Doctor–Patient Relationship in the Hospital Context,* 27 Vand. L. Rev. 243 (1974). *See also* Regan, *When Nursing Home Patients Complain: The Ombudsman or the Patient Advocate,* 65 Georgetown L. J. 691(1977); and Rehr & Ravich, "An Ombudsman in a Hospital: A Patient Representative," in Malick & Rehr, eds., *In the Patient's Interest* (New York: Prodist, 1981), at 68–79. AIDS patients are particularly vulnerable to discrimination. H. L. Dalton *et al., eds., AIDS and the Law* (New Haven, Conn.: Yale U. Press, 1987).
2. P. Starr, *The Social Transformation of American Medicine* (New York: Basic Books, 1982), at 389.
3. *Id.* at 391.

4. S. Terkel, *Working* (New York: Avon, 1972), at 647.
5. *See, e.g.,* Sun, "Patient Reps Have a Problem: Winning Acceptance in Hospitals," *Medical News,* Apr. 2, 1979, at 15; and Eisenberg, "Patient Representatives: Sometimes They Even Help Doctors," *Medical Economics,* Sept. 15, 1980, at 80; and Scheier, "A Dilemma a Day for Ombudsman," *American Medical News,* Feb. 5, 1988, at 21.
6. Gordon, "Insurers' Limits on Hospital Stay Cause Concern for New Mothers," Boston Globe, July 4, 1988, at 21, 23.
7. Survey of members of National Society of Patient Representatives, American Hospital Association, Feb. 1986. *See also* Mailick, "Models for Patient Representative Programs," in Mailick & Rehr, *supra* 1, at 85.
8. Oral presentation to the trustees, Sept. 26, 1973 (personal communication).
9. E. Kennedy, *In Critical Condition* (New York: Pocket Books, 1973), at 49.
10. Hanlan, "Notes of a Dying Professor," *Penn Gazette,* Feb. 1972, at 23 (his disease was eventually diagnosed as amyoatrophic lateral sclerosis [ALS], often called Lou Gehrig's disease); *see also* A. H. Malcolm, *This Far and No More* (New York: Times Books, 1987).
11. *Haggerty, Ms. Magazine,* Jan. 1973, at 16–17 *and see supra* note 6.
12. Some states require hospitals to have risk management systems as a way to improve the quality of care delivered in the hospital. Rhode Island's statute, for example, requires each hospital and its insurance carrier to set up a risk management system with the following components:

(1) an in-hospital grievance or complaint mechanism designed to process and resolve as promptly and effectively as possible grievances by patients or their representatives related to incidents, billing, inadequacies in treatment, and other factors known to influence medical malpractice claims and suits...

(2) ...the continuous collection of data by each hospital with respect to its negative health care outcomes (whether or not they give rise to claims), patient grievances, claims suits, professional liability premiums, [etc.]...

(3) ...medical care evaluation mechanisms....

(4) ...education programs for the hospital's staff personnel engaged in patient care activities dealing with patient safety, medical injury prevention, the legal aspects of patient care, problems of communication and rapport with patients, and other relevant factors known to influence malpractice claims and suits. (R.I. Gen. Laws sec. 23-17–24 [1979])

And see "Patient Advocacy Programs Seen Curbing Hospital Malpractice Losses,"*American Medical News,* Apr. 6, 1979, at 7; Rubin, "Medical Malpractice SuitsCan Be Avoided," *Hospitals, JAHA,* 1, 1978, at 86; J. E. Orlikoff & A. M. Vanagunas, *Malpractice Preven-*

tion and Liability Control for Hospitals, 2d ed. (Chicago: American Hospital Association, 1988).

13. *See* R. E. Cranford & A. E. Doudera, eds., *Institutional Ethics Committees and Health Care Decision Making* (Ann Arbor, Mich.: Health Administration Press, 1984).

14. M. Lear, *Heartsounds* (New York: Simon & Schuster, 1980); *see also* C. Ryan & K. M. Ryan, A Private Battle (New York: Simon & Schuster, 1979), at 47.

15. *Id.* at 142 43.

16. A model Patient Bill of Rights is set forth in appendix B.

17. *See* ch. X, "Medical Records."

18. *See* ch. VI, "Informed Consent."

Appendix A

Other Resources

Organizations that provide information or help

LISTS OF SELF-HELP PROFESSIONAL AND LICENSING ORGANIZATIONS

Inlander, C. B., L. S. Levin & E. Weiner. *Medicine on Trial.* New York: Prentice-Hall, 1988.

Contains addresses and phone numbers of all the state boards that license physicians, dentists, nurses, and hospitals—organizations with which consumers can lodge complaints against incompetent and negligent practitioners.

Madara, E. J. & A. Meese. *The Self-Help Sourcebook: Finding and Forming Mutual Aid Self Help Groups.* Denville, NJ: St. Clare's-Riverside Medical Center, 1986.

Available from New Jersey Self-Help Clearinghouse, St. Clares-Riverside Medical Center, Denville, NJ 07834. Lists and describes more than four hundred national self-help groups; most relate to health and medical conditions ($8).

People's Medical Society. *Dial 800 for Health.* Emmaus, Pa: People's Medical Society, 1987. Available from People's Medical Society, 14 East Minor Street, Emmaus, PA 18049; (215) 967-2136. A listing of toll-free phone numbers of more than 150 organizations that provide information about various medical conditions and local support groups ($4).

ORGANIZATIONS INVOLVED IN PATIENT RIGHTS

American Civil Liberties Union (ACLU)
132 West 43d Street
New York, NY 10036
(212)944-9800

National organization with state affiliates and local chapters that actively protects the constitutional rights of citizens. Most concerned in the health field with the right to privacy, equal protection, confidentiality, access to records, and equal access to care.

American Society of Law and Medicine
765 Commonwealth Avenue
Boston, MA 02215
(617)262-4990

Professional continuing education group that publishes the journals *Law, Medicine and Health Care* and *American Journal of Law & Medicine*, both good sources of current trends in health law.

Children in Hospitals, Inc.
31 Wilshire Park
Needham, MA 02192 (617)482-2915

Consumer organization dedicated to helping parents stay with and support their children during hospitalization. Publishes a newsletter.

Concern for Dying   and   Society for the Right to Die
250 W. 57th Street   250 W. 57th Street
New York, NY 10107   New York, NY 10107
(212)246-6980   (212)246-6973

Originally the same organization (Euthanasia Educational Council), these two organizations provide information and support for those interested in exercising their rights as patients, especially the right to refuse medical treatment. Concern for Dying developed the living will and has information on it; Right to Die publishes an annually updated collection of all state living will laws. These groups, which have a newsletter and provide assistance to individuals, merged in 1991 to form Choice in Dying.

The Hemlock Society
PO Box 11830
Eugene, OR 97440
(503) 342-5748

Promotes legislation to legalize physician-assisted suicide and distributes literature on this subject. A membership organization, Hemlock publishes a newsletter and a journal, the *Euthanasia Review*.

The Hastings Center
255 Elm Road
Briarcliff Manor, NY 10510
(914) 762-8500

The major bioethics "think tank" in the United States, the Hastings Center produces reports and guidelines on medical treatment and research issues and publishes the bimonthly *Hastings Center Report*, the leading publication on medical ethics. Other major medical ethics groups include the Kennedy Institute of Ethics at Georgetown University, the Medical Humanities Institute at University of Texas Medical Branch in Galveston, the Center for Biomedical Ethics at Case Western Reserve University, and the Center for Bioethics at the University of Minnesota.

Law, Medicine and Ethics Program
Boston U. Schools of Medicine and Public Health
80 E. Concord Street
Boston, MA 02118
(617) 638-4626

Does education, research, and advocacy in health law, with emphasis on patient rights, health care regulation, and medical ethics.

National Association of Parents and Professionals for
    Safe Alternatives in Childbirth
Route 1, Box 646
Marble Hill, MO 63764

Activist organization dedicated to individual control over childbirth and to improving childbirth. Supports home birth and midwifery. Organizes conferences and publishes a newsletter.

The National Hospice Organization
1901 North Forth Myer Drive (Suite 902)
Arlington, VA 22209
(703) 243-5900

Clearinghouse on information about hospices that publishes a national directory and will provide information to interested persons who write or call.

National Health Law Program

Main Office	*Branch Office*
2639b South La Cienega Boulevard	2025 M Street NW
Los Angeles, CA 90034	Suite 400
(213) 204-6010	Washington, DC 20036
	(202) 887-5310

Legal services backup center specializing in health law, Medicaid, and access issues for the poor; publishes a newsletter.

National Information Center for Handicapped Children and Youth
PO Box 1492
Washington, DC 20013
(703) 522-3332

Provides information on services for handicapped children and youth.

People's Medical Society
462 Walnut Street
Allentown, PA 18102
(215) 770-1670

Membership organization for consumers of medical services, it provides information on a variety of issues regarding patient rights and publishes a newsletter, books, and pamphlets. (See book list on next page for examples.)

Pike Institute
Boston University School of Law
765 Commonwealth Avenue
Boston, MA 02215
(617) 353-2910

Does work to promote rights of the handicapped and publishes the *Disability Advocates Bulletin.*

Planned Parenthood Federation of America, Inc.
810 Seventh Avenue
New York, NY 10019
(212)541-7800

National organization that runs more than seven hundred centers across the country that provide family planning services. Strong supporters of the "right to privacy" in family planning.

Public Citizen Health Research Group
2000 P Street NW (Suite 708)
Washington, DC 20036
(202)872-0320

Ralph Nader-affiliated consumer advocacy group concerned with issues of medical care, drug safety, medical device safety, physician competence, and consumer health care issues in general. Prepares many publications and offers testimony before Congress and regulatory agencies, as well as filing and participating in lawsuits on patient rights issues. Publishes the *Health Letter* monthly.

BOOKS ON PATIENT RIGHTS

Aspen System (Health Law Center). *The Hospital Law Manual.* Rockville, Md.: Aspen, 1988 (updated regularly).

A three-volume set of books for lawyers (and a separate three-volume set for administrators) that summarizes hospital law in general, as well as the law in each individual state. It provides additional references for almost every topic covered in the book and should be available in your local law school or bar association library. If not, the hospital administrator probably has a set. An excellent source for up-to-date activity by courts, Congress, regulatory agencies, and state legislatures on hospital law topics.

Boston Women's Health Book Collective. *The New Our Bodies, Ourselves.* New York: Simon & Schuster, 1984.

An outstanding reference book on women's health and how the medical establishment treats women. Written by and about women. The BWHBC is a nonprofit organization devoted to education about women and health located at 240-A Elm Street, Somerville, MA 02144. And for a woman's perspective on medical practice, see Michelle Harrison, *A Woman in Residence* (New York: Random House, 1982).

Consumer Guide. *Getting the Most from Social Security, Medicare, and Other Government Benefits.* New York: Signet, 1987.

This publication contains information on Medicare benefits, how to qualify for them, and your rights under Medicare. See also the

Social Security Systems publications on this subject, especially *Your Medicare Handbook*, available at your local Social Security office.

Davis, N. M. *Medical Abbreviations*, 4th ed. Huntington Valley, Pa.: Neil M. Davis Assoc., 1988.

Excellent guide to 5500 medical abbreviations.

Dalton, Harlon L. Scott Burris, and the Yale AIDS Law Project. *AIDS and the Law: A Guide for the Public*. New Haven, Conn.: Yale U. Press, 1987.

Most complete and authorative book available on the legal aspects of the AIDS epidemic, including issues of screening, health care, insurance, education, and employment.

Goldsmith, L. S., ed. *Medical Malpractice: Guide to Medical Issues*. New York: Matthew Bender, 1986.

A comprehensive guide designed to help lawyers prepare for malpractice litigation on *both* sides by presenting both the medical and the legal aspects of common malpractice occurrences in emergency medicine, anesthesiology, surgery, orthopedics, neurology and neurosurgery, and obstetrics and gynecology. (Available at law libraries.)

Gots, Ronald & Arthur Kaufman. *The People's Hospital Book*. 1978. Reprint. New York: Avon, 1981.

A useful description of what goes on in the hospital, including the "written and unwritten rules," and how preparing yourself for your hospital stay can help minimize the risk of injury and make the experience more understandable. Written from the physician's perspective.

Huttmann, Barbara. *The Patient's Advocate*. New York: Viking, 1981.

Useful information and strategy suggestions on how you can act as an advocate for your spouse, child, relative, or friend when they are hospitalized. Emphasizes the fact that in a hospital, "you need an advocate." Written from a nurse's pespective.

Inlander, C. B. & Ed Weiner. *Take This Book to the Hospital with You: A Consumer Guide to Surviving Your Hospital Stay*. Emmaus, Pa.: Rodale Press, 1985; New York: Warner Books,1985.

An aggressive "travel guide" to your hospital stay from a consumer point of view, containing practical information on how to get what you want during your hospital stay, and why you might not.

Nierenberg, J. & Florence Janovic. *The Hospital Experience: A Comprehensive Guide to Understanding and Participating in Your Own Care.* New York: Bobbs-Merrill, 1978.

Contains helpful descriptions of common tests, operations, and treatments; good aid in explaining these tests to nonphysicians. Also down-to-earth descriptions of hospital routines.

People's Medical Society. *Nursing Homes: How to Evaluate and Select a Nursing Home* (Emmaus, Pa.: People's Medical Society, 1983).

A thoughtful guide to help you choose and evaluate a nursing home. Includes a list of nursing home ombudsman offices and state nursing home licensure offices. Chapter 4, "Things to Consider When You Inspect a Nursing Home," is especially useful.

Rothenberg, Robert E. *The New American Medical Dictionary and Health Manual*, 5th ed. New York: New American Library 1988.

Contains more than 9000 definitions of medical terms, as well as a concise general health manual.

Scully, Thomas & Celia Scully. *Playing God: The New World of Medical Choices*. New York: Simon & Schuster, 1987.

An accessible description of many of the legal issues dealt with in this book from the perspective of a physician and his writer-wife who adopt a "medical ethics" perspective and provide many suggestions for further reading, as well as the names and addresses of organizations to contact for help with specific problems.

Sneider, Iris. *Patient Power: How to Have a Say During Your Hospital Stay.* White Hall, Va.: Betterway, 1986.

A social worker's perspective on how to help yourself during hospitalization.at the University of Washington, Seattle; 9) the Pacific Southwest Regional Medical Library Service at the Biomedical Library of the University of California, Los Angeles.

The statute provides that "qualified persons and organizations shall be entitled to free loan service." Although some of these regional libraries have attempted to restrict the meaning of qualified to other medical libraries, this definition is not consistent with the wording of the statute or its purpose. An excellent argument can be made for the proposition that any individual who can read and who has a desire to research a particular medical question is a "qualified" user under the Act and should be given not only access to the library but free loan privileges as well. This proposition has not yet been argued in court (*see* 42 C.F.R. 59a.31 *et seq.*).

Once you obtain access to the medical library, your main source of information will be the card catalog for books and medical indexes for periodicals. The major index is *Index Medicus,* which is published by the National Institutes of Health (NIH) and indexes about 3000 worldwide medical periodicals under approximately 8000 subject headings. These subject headings are constantly being updated and appear each year as part II of the January issue designated "Medical Subject Headings." Listings are also grouped by the author's last name. In addition to the yearly indexes, there are also *cumulative* edition and an *abridged* edition of *Index Medicus.* There are also approximately 15 other more specialized indexes to the medical and biological literature.

Finally, computer searches can locate a complete listing of all articals published over a given period of time on a particular subject matter. Further information on ordering these searches and their cost can be obtained from the reference librarian of your local medical library.

For more specific information on searching the medical literature, *see* Beatty, *Searching the Literature Comes before Writing the Literature,* 79 Ann. Internal Med. 917-24 (1973).

Appendix B
Patient Bill of Rights Act

SECTION 1. PREAMBLE

The rights established under this act shall apply to every patient or resident in a general hospital, outpatient clinic, health maintenance organization, nursing home, state hospital, or other health care facility licensed, or subject to licensing, by the [state licensing authority].

This act shall not prohibit a health facility or agency from establishing and recognizing other additional patient rights and shall not limit the exercise and enforcement of other rights provided by law.

SECTION 2. DEFINITIONS

2.1) "Patient" means any individual receiving long- or short-term inpatient care, emergency care, or outpatient care, and residents of long-term care facilities.

2.2) "Facility" means hospitals, clinics, health maintenance organizations, nursing homes, and other health care facilities licensed, or subject to licensing, by the [state licensing authority].

2.3) "Physician" means any staff, attending, visiting, resident, or intern physician who cares for any patient in a licensed facility.

2.4) "Patient Rights Advocate" means an individual whose primary job is to help patients exercise their rights.

SECTION 3. FACILITY DUTIES TO PROMOTE EXERCISE OF PATIENT RIGHTS

Each facility has the duty:

3.1) To provide each patient with a written copy of the rights protected in this act and other patient rights adopted by the facility on acceptance of the patient for care;

3.2) To post the rights outlined in paragraph 3.1 conspicuously in every patient room and at every entrance;

3.3) To distribute and post this list of rights in English and any other language spoken by 5 percent or more of the facility's patients;

3.4) To provide each patient with 24-hour-a-day access to a patient rights advocate, who may act on behalf of the patient, with the patient's consent, to assert or protect the rights set forth in this act;

3.5) To encourage and assist each patient to understand and exercise his or her rights and to voice grievances and recommend changes in policies and services to facility staff and outside representatives of their choice, free from facility restraint, interference, coercion, discrimination, or reprisal;

3.6) To inform each patient of all rules governing patient or resident conduct and responsibilities.

SECTION 4. FACILITY DUTIES TO PROMOTE PATIENT DIGNITY

Each facility has the duty:

4.1) To provide each patient the opportunity for access to people outside the health care facility by means of visitors, mail, and the telephone;

4.2) To provide patients the opportunity to meet with clergy or other religious advisers;

4.3) To provide patients the opportunity to associate and communicate privately with persons of their choice and to send and receive personal mail unopened;

4.4) To provide each patient the opportunity for privacy during visits;

4.5) To permit patients to share a room with their spouses, if both spouses are residents of the facility;

4.6) To provide parents and guardians of minor children, patients, and relatives of terminally ill patients the opportunity to stay with these patients twenty-four hours a day;

4.7) To provide each child-bearing woman the opportunity to have a spouse or person of her choice with her during labor, delivery, and postpartum and the opportunity to have her baby stay with her after birth;

4.8) To allow patients to retain and use personal clothing and possessions as space permits;

4.9) To ensure that patients are not required to perform services for the facility that are not included for therapeutic purposes in their plan of care;

4.10) To provide a prompt, reasonable response to patient requests for services or information;

4.11) To care for each patient free from chemical and physical restraints, except in emergencies or as authorized in writing by the patient's physician for a specified and limited period of time when necessary to protect the patient from injury to himself or others;

4.12) To provide each patient who does not speak English, or who is deaf and requires an interpreter, with access to an interpreter.

Section 5. Facility Duties to Render Services

Each facility has the duty:

5.1) To provide each patient with considerate, respectful, patient care by competent personnel;

5.2) To provide high-quality care and high professional standards that are maintained and reviewed regularly by the hospital and its professional staffs;

5.3) To provide prompt treatment in an emergency without discrimination on account of economic status or source of payment and without delaying treatment for purposes of ascertainment of the source of payment;

5.4) To provide medical and nursing services without discrimination based on race, color, religion, sex, sexual preference, handicap, disease, marital status, national origin, or source of payment;

5.5) To provide care based on an individual treatment plan designed or approved specifically for that patient by the physician and to provide drugs, tests, procedures, or treatments only if specifically indicated for that patient and not to subject any patient to routine facility procedures unnecessary to their care;

5.6) To provide each patient with an itemized and detailed explanation of the total bill for services rendered in the facility regardless of source of payment;

5.7) To provide each patient with competent counseling and assistance to help the patient obtain financial assistance from public or private sources to meet the expense of services received in the institution;

5.8) To provide each patient with timely prior notice of the termination of eligibility for reimbursement by any third-party payer for the expense of his or her care;

5.9) To provide each patient with continuity of care within each facility, responding to patient needs by immediate care and treatment or by referral at time of discharge.

Section 6. Facility and Physician Duties to Provide Information and Ensure Informed Consent

Each facility and physician within each facility has the duty:

6.1) To provide each patient with the opportunity for informed participation in all decisions involving that patient's health care program;

6.2) To provide each patient clear, concise, complete, and current explanation of all proposed procedures in terms and language the patient can reasonably be expected to understand, including alternatives and success rates, and the possibilities of any risk of death or serious bodily harm, problems related to recuperation, and probability of success, and what the physician means by success, and the cost, and not to subject any patient to any procedure without receiving voluntary, competent, and understanding consent documented in writing and signed by the patient (this requirement does not apply to procedures that carry no risk of death or serious harm if the nature and likely consequences of the procedure are common knowledge, or if the patient states he or she does not want to be so informed);

6.3) To provide each patient with a clear, complete, and accurate evaluation of his or her condition and prognosis without treatment, before asking that patient to consent to any test or procedure, and to describe the facility's and physician's experience with that test or procedure, including the complication rate and whether the facility and/or physician have ever performed the procedure;

6.4) To inform each potential patient what recommended procedures are research or experimental in nature, what alternative protocols are available for that patient's care within the facility, and if known, what alternatives are available in the community;

6.5) To inform the patient if a proposed test or procedure is designed primarily or exclusively for educational purposes rather than for his or her direct personal benefit;

6.6) To ensure that all personnel wear name tags and to inform each patient of the identity and professional status of all those providing service, to ensure that all personnel introduce themselves to the patient, state their professional status and position within the facility, and explain their role in the health care of the patient;

6.7) To provide each patient, on request, the opportunity to inspect that patient's medical records and, on payment of reasonable copying expenses, to receive a complete copy of the records;

6.8) To inform each patient of the name of the physician responsible for the patient's care;

6.9) To inform each patient, on request, the opportunity to discuss his or her condition with another physician or a consultant specialist at the patient's expense;

6.10) To inform, on request, each patient of the relationship of the facility to other health care and related institutions insofar as that relationship relates to care or treatment.

SECTION 7. FACILITY AND PHYSICIAN DUTIES TO ENSURE CONFIDENTIALITY AND PRIVACY

Each facility and physician in each facility has the duty:

7.1) To prepare and manage all communications and records pertaining to each patient's care or personal information in a confidential and discreet manner;

7.2) To conduct case discussion, consultation, examination, and treatment, or other rendering of care in a confidential and discreet manner;

7.3) To establish policies limiting disclosure of or access to the confidential information described above to physicians, residents, interns, medical students, researchers, nurses, other hospital personnel, and other patients only by permission of the patient or as necessary for patient care.

SECTION 8. FACILITY AND PHYSICIAN DUTIES TO ENSURE A PATIENT'S RIGHT TO REFUSE TREATMENT

Each facility and physician in each facility has the duty:

8.1) To respect the refusal of any competent adult patient to initial or continued use of any drug, test, procedure, or treatment, even

if that patient has previously given informed consent, and even
if such refusal will lead to the patient's death;

8.2) To provide any patient refusing treatment with concise, clear
information concerning the medical consequences of refusal;

8.3) To ensure medical, nursing, psychiatric, psychological, or other
care without penalty or harassment for patients who refuse:

a) to be examined, observed, or treated by students or other
facility staff;

b) to serve as subjects in experimental research;

c) to consent to receive any drug, test, procedure, or treatment.

8.4) To honor the provisions of a patient's written declaration concerning care after the patient has become incompetent and the
designation of another person to make health care decisions
on the patient's behalf.

Section 9. Facility and Physician Duties Relating to Patient Discharge, Transfer, or Departure Against Medical Advice

Each facility and physician has the duty:

9.1) To provide each patient the opportunity to leave the facility regardless of physical condition or financial status, though patients may be asked to sign a release stating that they are leaving against the medical judgment of the facility or physician;

9.2) To discharge or transfer patients only for medical reasons that
improve their welfare;

9.3) To inform patients of their discharge at least one day prior to
discharge and to provide them the opportunity to notify a person
of their choice regarding their impending discharge;

9.4) To provide each patient, prior to another facility, with a complete explanation of the desirability and need for the transfer
and the other facility's acceptance of the patient for transfer;

9.5) To provide any patient refusing transfer with the opportunity to
obtain a second opinion from a physician or other consultant
on the medical desirability of transfer;

9.6) To provide any patient with assistance in planning for their
discharge, transfer, or withdrawal from the facility, including
information concerning their continuing health needs and alternatives for meeting those needs.

SECTION 10. ENFORCEMENT PROCEDURES

10.1) Each facility shall adopt a patient rights enforcement plan describing procedures to enforce patient rights and to ensure effective and fair investigation of possible violations:

   a) Establish a system to identify, record, and investigate formal written and oral complaints;

   b) Keep complaint records on file for at least ten years and make them available to the department of public health upon request;

   c) Investigate and resolve formal complaints in a timely manner and notify patients of expected time frame;

   d) Develop disciplinary and remedial education procedures for members of the hospital and medical staff who consistently cause patient relationship problems;

   e) Hire at least one patient rights advocate with authority to obtain patient records, present patient concerns to responsible physicians and nursing staff and other personnel directly involved in patient care, delay discharge, investigate and report on facility policies impacting on patient rights, refer patients to internal and external support groups for specific problems, educate patients and staff to the provisions of this act and the hospital's policies respecting it.

10.2) Each facility shall establish these and other enforcement procedures subject to regulations promulgated by the [licensing authority] pursuant to this act, and shall submit a report on these enforcement procedures to the [licensing authority] annually. The frequency, form, and content of each report shall be prescribed by regulation of the state [licensing authority].

10.3) Each facility shall submit its proposed patient rights enforcement plan to the [licensing authority] for approval prior to implementing the plan.

10.4) The [licensing authority] shall, within 120 days from the effective date of this act, establish by regulation:

   a) Minimum standards and procedural requirements for facility complaint mechanism;

   b) A list of patient complaints that may be processed through the complaint mechanism;

   c) The form and manner in which facilities must inform patients of their rights;

d) A schedule of fines and other disciplinary measures imposed for the failure of a facility or physician to comply with the provisions of this section;

e) The frequency, form, and content of facility patient rights enforcement reports to the [licensing authority]; and

f) Policies for referring facilities or physicians who violate the provisions of this section to the [licensing authority] medical society, or JCAH as appropriate;

g) An office of health rights advocates within the state [licensing authority] of health to receive, investigate, resolve, and record patient complaints received directly from the public.

10.5) Any person whose rights are violated under this act may bring, in addition to any other action allowed by law or regulation, a civil action under the state's consumer protection act.

COMMENT

This model bill is specific and enforceable. It was drafted from a patient rights perspective, but represents a realistic appraisal of what rights are achievable in modern health care facilities and enforceable in the courts: *The bill assigns specific duties to facilities and physicians and provides a comprehensive enforcement system on both hospital and state levels.*

The model bill focuses on hospital patient rights, but also includes general provisions applicable to nursing home and mental health patients. An ideal regulatory scheme would augment this bill with specific regulations covering nursing homes and mental health facilities. With respect to nursing homes, the Massachusetts Attorney General Regulations offer a useful model.[1] New York State Regulations outline some specific areas important to mental health patients, including standards on seclusion, restraint, sterilization, family planning, emergency treatment, and psychosurgery.[2]

The model bill applies only to licensed facilities and the physicians who practice within them. Facilities are charged with duties to provide services, promote patient rights, and respect patient dignity. Both physicians and facilities must ensure confidentiality and privacy, provide adequate information, and ensure informed consent. Facilities and physicians must further guard a patient's right to refuse treatment and protect a patient's rights during discharge, transfer, or departure against medical advice.

Finally, a detailed enforcement procedure requires health facilities to adopt their own patient rights enforcement plans monitored and regulated by the state department of health. The bill further mandates the health department to establish complaint mechanism guidelines, a schedule of disciplinary measures, and a special office to handle complaints received directly from the public.

As of 1990, the following states had enacted some version of a Patient Bill of Rights applicable to hospital patients by statute, regulation, or resolution: California[3], Colorado,[4] Illinois,[5] Kentucky, [6] Maryland,[7] Massachusetts,[8] Michigan,[9] Minnesota,[10] Pennsylvania,[11] New York,[12] and Rhode Island.[13]

NOTES

1. 900 C.M.R. 4.00–4.09 (1975) (Attorney General Regulations) comprise one part of a three-pronged state regulatory scheme governing nursing home patient rights. The remaining two sections are Department of Public Health Regulations on Nursing Homes, 105 C.M.R. 150.000 150.020 (1977); and the state patient rights act, Mass. Gen. Laws Ann. ch. 111, sec. 70 (E) (West 1979). For an analysis of the Massachusetts scheme, see *Legislative Comment, Nursing Home Patients' Rights in Massachusetts: Current Protection and Recommendation for Improving,* 6 Am. J. Law & Med. 315 (1980). The fight for reasonable federal rules to protect nursing home patients continues. *See* Tolchin, "New Rules Issued on Patients' Care in Nursing Homes," New York Times, Jan. 31, 1989, at 1.
2. N.Y. Code of Rules and Regulations, 14 NYCM *et seq.*
3. Cal. Admin. Code tit. 22, R.70707.
4. Colo. Rev. Stat. sec. 25-1-121(1982).
5. Ill. Ann. Stat. ch. 1111/2, sec. 5401–5404 (Smith-Hurd Supp. 1985).
6. House Resolution 151 (Mar. 17, 1986) (86 RS BR 2574).
7. Md. Health Gen. Code Am. sec. 19-342 (1982).
8. Mass. Gen. Laws Ann. ch. 111, sec. 70 (E) (West 1979).
9. Mich. Comp. Law Ann. sec. 333.20201 (West 1979).
10. Minn. Stat. Ann. sec. 144.651 (West 1980).
11. Pa. Admin. Code 103.22 (1980).
12. N.Y. Admin. Code tit. 10, sec. 405.25 (1975).
13. R. I. Gen. Laws sec. 23-17-19.1 (1985).

Appendix C
Health Care Proxy*

I, _____, residing at
    (name of principal)

_____, Massachusetts
    (street)   (city)

do hereby appoint: _____, residing at
    (name)

_____ as my **Health Care Agent**
  (street)   (city)   (phone)

with the authority to make all health care decisions for me, including decisions about life-sustaining treatment, subject to any limitations I state below, if I am unable to make health care decisions myself. My Agent's authority becomes effective if my attending physician determines in writing that I lack the capacity to make or to communicate health care decisions myself. My Agent is then to have the same authority to make health care decisions as I would if I had the capacity to make them EXCEPT (here list the limitaions, *if any*, you wish to place on your Agent's authority):

I direct my health care Agent to make decisions based on my agent's assessment of my personal wishes. If my personal wishes are unknown, my agent is to make decisions based on my agent's assessment of my best interests. Photocopies of this Health Care Proxy shall have the same force and effect of the original.

_____(signed)
            (principal)

Witness statement

We, the undersigned witnesses, each declare that we know the identity of the person who signed this Health Care Proxy, that the person appears to be at least 18 years of age, of sound mind, and under no constraint or undue influence. Neither of us is named the health care agent in this document. Both of us witnessed the signature by the person who signed this Health Care Proxy, or witnessed it signed at the person's direction, in our presence this _____ day of _____ 199__.

Witness One: _____ Witness Two: _____
Name (print): _____ Name (print): _____

_____   _____
_____   _____

*Massachusetts. *See* Annas, G.J., The Health Care Proxy and the Living Will, 324 *New Engl. J. Med. 1210 (1991)*.

Sample Living Will and Durable Power of Attorney for Health Care Decisions

LIVING WILL AND APPOINTMENT OF SURROGATE DECISION MAKER

To My Family, My Physician, My Lawyer, and All Others Whom It May Concern

Living Will

Death is as much a reality as birth, growth, and aging—it is the one certainty of life. In anticipation of the decisions that may have to be made about my own dying and as an expression of my right to refuse treatment, I, _____, being of sound mind, make this statement of my wishes and instructions concerning treatment.

By means of this document, which I intend to be legally binding, I direct my physician and other care providers, my family, and any surrogate designated by me or appointed by a court to carry out my wishes. If I become unable, by reason of physical or mental incapacity, to make decisions about my medical care, let this document provide the guidance and authority needed to make any and all such decisions.

If I am permanently unconscious or there is no reasonable expectation of my recovery from a seriously incapacitating or lethal illness or condition, I do not wish to be kept alive by artificial means. I request that I be given all care necessary to keep me comfortable and free of pain, even if pain-relieving medications may hasten my death, and I direct that no life-sustaining treatment and no artificial nutrition be provided except as I or my surrogate specifically authorizes.

This request may appear to place a heavy responsibility upon you, but by making this decision in accordance with my strong convictions, I intend to ease that burden. I am acting after careful consideration and with understanding of the consequences of your carrying out my wishes.

(optional specific provisions to be made in this space)

Durable Power of Attorney for Health Care Decisions
(delete this section if inapplicable)

To effect my wishes, I designate_____, residing at _____, as my health care surrogate—that is, my attorney-in-fact regarding any and all health care decisions to be made for me, including the decision to refuse life-sustaining treatment—if I am unable to make such decisions myself. If this person is unavailable, I designate_____, residing at _____, as my alternate surrogate. This power shall remain effective during, and not be affected by, my subsequent illness, disability, or incapacity. My surrogate shall have authority to interpret my Living Will and shall make decisions about my health care as specified in my instructions or, when my wishes are not clear, as the surrogate believes to be in my best interests. I release and agree to hold harmless my health care surrogate from any and all claims whatsoever arising from decisions made in good faith in the exercise of this power.

    I sign this document knowingly, voluntarily, and after careful deliberation, this_____ day of_____, 19____

Address_____ (s_____
_____   (signature)
_____

I do hereby certify that the within   Witness (s)_____
document was executed and acknowl- _____
edged before me by the principal this Printed Name_____
day of_____19___ . Address_____
_____

_____ Witness (s)_____
    Notary Public Printed Name _____
     (seal) Address _____
_____

Source: Concern for Dying, 250 W. 57th St., New York, NY 10107.

Statements of Health Care Agent and Alternate (OPTIONAL)

Health Care Agent: I have been named by the Principal as the Principal's Health Care Agent by this Health Care Proxy. I have read this document carefully, and have personally discussed with the Principal his/her health care wishes at a time of possible incapacity. I know the Principal and accept this appointment freely. I am not an operator, administrator or employee of a hospital, clinic, nursing home, rest home, Soldiers Home or other health facility where the Principal is presently a patient or resident or has applied for admission. Or if I am a person so described, I am also related to the Principal by blood, marriage, or adoption. If called upon and to the best of my ability, I will try to carry out the Principal's wishes.

(Signature of **Health Care Agent**) _____

Alternate: I have been named by the Principal as the Principal's Alternate by this Health Care Proxy. I have read this document carefully, and have personally discussed with the Principal his/her health care wishes at a time of possible incapacity. I know the Principal and accept this appointment freely. I am not an operator, administrator or employee of a hospital, clinic, nursing home, rest home, Soldiers Home or other health facility where the Principal is presently a patient or resident or has applied for admission. Or if I am a person so described, I am also related to the Principal by blood, marriage, or adoption. If called upon and to the best of my ability, I will try to carry out the Principal's wishes.

(Signature of **Alternate**)_____

MASSACHUSETTS HEALTH CARE PROXY

Information, Instructions, and Form

What does the Health Care Proxy Law allow?

The **Health Care Proxy** is a simple legal document that allows you to name someone you know and trust to make health care decisions for you if, for any reason and at any time, you become unable to make or communicate those decisions. It is an important document, however, because it concerns not only the choices you make about your health care, but also the relationships you have with your physician, family, and others who may be involved with your care. Read this and follow the instructions to ensure that your wishes are honored.

Under the Health Care Proxy Law (Massachusetts General Laws, Chapter 201D), any competent adult 18 years of age or over may use this form to appoint a Health Care Agent. You (the "Principal") can appoint anyone EXCEPT the administrator, operator, or employee of a health care facility such as a hospital or nursing home where you are a patient or resident UNLESS that person is also related to you by blood, marriage, or adoption.

What can my Agent do?

Your Agent will make decisions about your health care only when you are, for some reason, unable to do that yourself. This means that your Agent can act for you if you are temporarily unconscious, in a coma, or have some other condition in which you cannot make or communicate health care decisions. Your Agent cannot act for you until your doctor determines, in writing, that you lack the ability to make health care decisions. Your doctor will tell you of this if there is any sign that you would understand it.

Acting with your authority, your Agent can make any health care decision that you could, if you were able. If you give your Agent full authority to act for you, he or she can consent to or refuse any medical treatment, including treatment that could keep you alive.

Your Agent will make decisions for you only after talking with your doctor or health care provider, and after fully considering all the options regarding diagnosis, prognosis, and treatment of your illness or condition. Your Agent has the legal right to get any information, including confidential medical information, necessary to make informed decisions for you.

Your Agent will make health care decisions for you according to your wishes or according to his/her assessment of your wishes, including your religious or moral beliefs. You may wish to talk first with your doctor, religious advisor, or other people before giving instructions to your Agent. It is very important that you talk with your Agent so that he or she knows what is important to you. If your Agent does not know what your wishes would be in a particular situation, your Agent will decide based on what he or she thinks would be in your best interests. After your doctor has determined that you lack the ability to make health care decisions, if you still object to any decision made by your Agent, your own decisions will be honored unless a Court determines that you lack capacity to make health care decisions.

Your Agent's decisions will have the same authority as yours would, if you were able, and will be honored over those of any other person, except for any limitation you yourself made, or except for a Court Order specifically overriding the Proxy.

How do I fill out the form?

1 At the top of the form, print your full name and address. Print the name, address, and phone number of the person you choose as your Health Care Agent. (**Optional:** If you think your Agent might not be available at any future time, you may name a second person as an Alternate. Your Alternate will be called if your Agent is unwilling or unable to serve.)

2 Setting limits on your Agent's authority might make it difficult for your Agent to act for you in an unexpected situation. If you want your Agent to have full authority to act for you, leave the limitations space blank. If, however, you want to limit the kinds of decisions you would want your Agent or Alternate to make for you, include them in the blank.

3 **BEFORE** you sign, be sure you have two adults present who can witness you signing the document. The only people who cannot serve as witnesses are your Agent and Alternate. Then sign the document yourself. (Or, if you are physically unable, have someone else sign at your direction. The person who signs your name for you should put his/her name and address in the spaces provided.)

4 Have your witnesses fill in the date, sign their names and print their names and addresses.

5 **OPTIONAL:** On the back of the form are statements to be signed by your Agent and any Alternate. This is not required by law, but is recommended to ensure that you have talked with the person or persons who may have to make important decisions about your care and that each of them realizes the importance of the task they may have to do.

Who should have the original and copies?

After you have filled in the form, remove this information page and make at least four photocopies of the form. Keep the original yourself where it can be found easily (not in your safe deposit box). Give one copy to your doctor who will put it in your medical record. Give copies to your Agent and any Alternate. You can give additional copies to family members, your clergy and/or lawyer, and other people who may be involved in your health care decisionmaking.

How can I revoke or cancel the document?

Your Health Care Proxy is revoked when any of the following four things happen:

1. You sign another Health Care Proxy later on.
2. You legally separate from or divorce your spouse and your spouse is named in the Proxy as your Agent.
3. You notify your Agent, your doctor, or other health care provider, orally or in writing, that you want to revoke your Health Care Proxy.
4. You do anything else that clearly shows you want to revoke the Proxy, for example, tearing up or destroying the Proxy, crossing it out, telling other people, etc.

Appendix D

Right to Refuse Treatment Act

Section 1. Definitions

"Competent person" shall mean an individual who is able to understand and appreciate the nature and consequences of a decision to accept or refuse treatment.

"Declaration" shall mean a written statement executed according to the provisions of this Act, which sets forth the declarant's intentions with respect to medical procedures, treatment or nontreatment, and may include the declarant's intentions concerning palliative care.

"Declarant" shall mean an individual who executes a declaration under the provisions of this Act.

"Health care provider" shall mean a person, facility, or institution licensed or authorized to provide health care.

"Incompetent person" shall mean a person who is unable to understand and appreciate the nature and consequences of a decision to accept or refuse treatment.

"Medical procedure or treatment" shall mean any action taken by a physician or health care provider designed to diagnose, assess, or treat a disease, illness, or injury. These include, but are not limited to, surgery, drugs, transfusions, mechanical ventilation, dialysis, resuscitation, artificial feeding, and any other medical act designed for diagnosis, assessment, or treatment.

"Palliative care" shall mean any measure taken by a physician or health care provider designed primarily to maintain the patient's comfort. These include, but are not limited to, sedatives and pain-killing drugs; nonartificial oral feeding; suction; hydration; and hygienic care.

"Physician" shall mean any physician responsible for the declarants care.

Section 2

A competent person has the right to refuse any medical procedure or treatment and any palliative care measure.

SECTION 3

A competent person may execute a declaration directing the withholding or withdrawal of any medical procedure or treatment or any palliative care measure, that is in use or may be used in the future in the person's medical care or treatment, even if continuance of the medical procedure or treatment could prevent or postpone the person's death from being caused by the person's disease, illness, or injury. The declaration shall be in writing, dated, and signed by the declarant in the presence of two adult witnesses. The two witnesses must sign the declaration, and by their signatures indicate they believe the declarant's execution of the declaration was understanding and voluntary.

SECTION 4

If a person is unable to sign a declaration because of a physical impairment, the person may execute a declaration by communicating agreement after the declaration has been read to the person in the presence of the two adult witnesses. The two witnesses must sign the declaration, and by their signatures indicate the person is physically impaired so as to be unable to sign the declaration, that the person understands the declaration's terms, and that the person voluntarily agrees to the terms of the declaration.

SECTION 5

A declarant shall have the right to appoint in the declaration a person authorized to order the administration, withholding, or withdrawal of medical procedures and treatment in the event that the declarant becomes incompetent. A person so authorized shall have the power to enforce the provisions of the declaration and shall be bound to exercise this authority consistent with the declaration and the authorized person's best judgment as to the actual desires and preferences of the declarant. No palliative care measure may be withheld by an authorized person unless explicitly provided for in the declaration. Physicians and health care providers caring for incompetent declarants shall provide such authorized person all medical information which would be available to the declarant if the declarant were competent.

SECTION 6

Any declarant may revoke a declaration by destroying or defacing it, executing a written revocation, making an oral revocation, or by any other act evidencing the declarant's specific intent to revoke the declaration.

SECTION 7

A competent person who orders the withholding or withdrawal of treatment shall receive appropriate palliative care unless it is expressly stated by the person orally or through a declaration that the person refuses palliative care.

SECTION 8

This Act shall not impair or supersede a person's legal right to direct the withholding or withdrawal of medical treatment or procedures in any other manner recognized by law.

SECTION 9

No person shall require anyone to execute a declaration as a condition of enrollment, continuation, or receipt of benefits for disability, life, health, or any other type of insurance. The withdrawal or withholding of medical procedures or treatment pursuant to the provisions of the Act shall not affect the validity of any insurance policy and shall not constitute suicide.

SECTION 10

This Act shall create no presumption concerning the intention of a person who has failed to execute a declaration. The fact that a person has failed to execute a declaration shall not constitute evidence of that person's intent concerning treatment or nontreatment.

SECTION 11

A declaration made pursuant to this Act, an oral refusal by a person, or a refusal of medical procedures or treatment through an authorized person, shall be binding on all physicians and health care providers caring for the declarant.

SECTION 12

A physician who fails to comply with a written or oral declaration and to make necessary arrangements to transfer the declarant shall be subject to civil liability and professional disciplinary action, including license revocation or suspension. When acting in good faith to effectuate the terms of a declaration or when following the direction of an authorized person appointed in a declaration or when following the direction of an authorized person appointed in a declaration under Section 5, no physician or health care provider shall be liable in any civil, criminal, or administrative action for withholding or withdrawing any medical procedure, treatment, or palliative care measure. When acting in good faith, no witness to a declaration, or person authorized to make treatment decisions under Section 5, shall be liable in any civil, criminal, or administrative action.

SECTION 13

A person found guilty of willfully concealing a declaration, or falsifying or forging a revocation of a declaration, shall be subject to criminal prosecution for a misdemeanor [the class or type of misdemeanor is left to the determination of individual state legislatures].

SECTION 14

Any person who falsifies or forges a declaration, or who willfully conceals or withholds information concerning the revocation of a declaration, with the intent to cause a withholding or withdrawal of life-sustaining procedures from a person, and who thereby causes life-sustaining procedures to be withheld or withdrawn and death to be hastened, shall be subject to criminal prosecution for a felony [the class or type of felony is left to the determination of individual state legislatures].

SECTION 15

If any provision or application of this Act is held invalid, this invalidity shall not affect other provisions or applications of the Act that can be given effect without the invalid provision or application, and to this end the provisions of this Act are severable.

COMMENT

This Act was originally drafted by the Legal Advisers Committee of Concern for Dying (250 W. 57th St., New York, NY 10107) and first published in 73 American Journal of Public Health 918 (1983); and as an appendix in the Report of the President's Commission for the Study of Ethical Problems in Medicine, *Deciding to Forego Life-Sustaining Treatment* (1983), at 428–31. The members of the drafting committee were: George J. Annas (chair), J. Dinsmore Adams, Leonard H. Glantz, Jane Greenlaw, Jay Healey, Barbara Katz, John Robertson, Richard Stanley Scott, Margaret Somerville, C. Dickerman Williams, and Ken Wing. The Act was rejected by a vote of 249 in the Massachusetts senate in the fall of 1986. The Republican leader, John Parker, opposed the bill, saying it was ahead of its time. For a more complete discussion of the Act, and a comparison between it and the Uniform Rights of the Terminally Ill Act, *see* Annas & Glantz, *The Right of Elderly Patients to Refuse Life-Sustaining Treatment,* 64 Milbank Quarterly 95 (Supp. 2, 1986); and ch. 3, "Legal Issues," in Congress of the United States, Office of Technology Assessment, *Life-Sustaining Technologies and the Elderly* (July, 1987).

Appendix E

How to Use Law and Medical Libraries

Few practicing lawyers and physicians know how to properly use either the law library or the medical library. The purpose of this brief appendix is not to teach you something few professionals have taken the time to understand, but to give you enough information so that if you want to find something out about a particular legal issue or medical condition, you will be able to do so. The first rule of research in any unfamiliar library is to *ask the reference librarian for assistance*. Do not, however, request such assistance until you have narrowed down your topic to the precise question that interests you.

THE LAW LIBRARY

All law schools have substantial libraries, as do many local bar associations. To obtain admission to the law library of your local law school, you may need special permission from the school or the assistance of a law student. Once inside, you will discover the principal problem with writing about "the law" in the United States: Each of the fifty states has its own court system and legislature, and therefore, each has its own set of statutes and case reporters. Superimposed on this structure is a system of federal district courts and federal appeals courts. Over them all is the US Supreme Court. In addition, there are not only the statutes of the United States (which are in a set of books called *US Code Annotated,* abbreviated USC) but also the regulations adopted under federal statutes by the agencies of the executive branch of government, such as the FDA, HHS (formerly HEW), and the Department of Agriculture (set forth in another set of books called the *Code of Federal Regulations,* or CFR). The legal literature you are most likely to be interested in locating is *statutes, court decisions* (case law), *regulations, legal periodicals, and legal encyclopedias.*

Statutes. Statutes are arranged by state, each having its own set of books (usually 50 to 100 volumes). These statutes are usually arranged by subject matter, and each provision of the statute has a number. If you know the number, the task of locating the statute is not difficult. If you do not, look up the subject matter in the index to the set. Note that since new statutes are passed each year, these volumes have "pocket parts" at the back in which up-to-date material is kept.

Court decisions. If you are looking for a particular case, you probably have the case name and citation. For example, the *Darling* case is properly cited as:

Darling v. Charleston Community Memorial Hospital, Ill. 2d 326, 211 N.E.2d 253 (1965), cert. denied, 383 U.S. 946 (1966)

To locate this case in the library, find either the set of books for the state of Illinois or the set labeled Northeastern Reporters (abbreviated NE). In the Illinois (Ill.) set, locate the second series (2d.). (Most states begin renumbering their reports after volume 200 or 300, but some do not). Within that set, find volume 33. The *Darling* case begins in volume 33 at page 326 (or in the N.E.2d series in volume 211 at page 253). The final part of the citation refers to the fact that the US Supreme Court refused to hear an appeal of the case. This refusal can be found in the United States Supreme Court Reports (US) volume 383, at page 946. The year in parentheses following the case is the year it was decided.

Regulations. Regulations are promulgated under statutory authority. Federal regulations are collected in the *Code of Federal Regulations*, parts of which are changed almost daily. These changes are reported in the Federal Register (Fed. Reg. or FR). State regulations are found in various, nonuniform state publications. In most states, locating regulations is very difficult.

Legal encyclopedias. Corpus Juris Secundum (C.J.S., the one Perry Mason used) and *American Jurisprudence* (Am. Jur.) are usually located side by side in a conspicuous part of the library. They are legal encyclopedias, general works on various aspects of the law, arranged alphabetically by subject matter.

Legal periodicals. The other type of legal material cited in many of the footnotes in this book is legal periodicals. These are usually published at individual law schools and are named after these law schools. For example, "54 B.U L. Rev." is the fifty-fourth volume of the set of Boston University's Law Review. As with case citations, the volume number appears *before* the name of the journal, the page number immediately after the journal's name. There is a rather unsophisticated and spotty index to these periodicals called the *Guide to Legal Periodicals* that is arranged by subject matter and author's last name.

For detailed information on legal research, *see* M. Price & H. Bitner, *Effective Legal Research,* 4th ed. (Boston, Mass.: Little Brown, 1979); J. Jacobstein & R. Mersky, *Pollock's Fundamentals of Legal Research,* 4th ed., (Brooklyn, NY: Foundation Press, 1973). For a more detailed treatment of the various aspects of lawmaking, see R. Covington *et al., Cases and Materials for a Course on Legal Methods* (Mineola, N.Y.: Foundation Press, 1969).

THE MEDICAL LIBRARY

In many ways, the medical library is a much easier library to deal with, since the "laws" of anatomy and pharmacology do not vary from state to state. Medical libraries are generally divided into three sections relevant to the consumer-patient: the *reference* section, *books*, and *periodicals.* The reference section will contain such materials as the list of medical specialists. The book section will contain such things as textbooks on various parts of the body and various medical specialities. The heart of the medical library, however, is its periodical collection, because it is here that new research and reports on currently used procedures are first published. A good medical library will receive from 2000 to 6000 periodicals. The largest collection of medical literature in this country is located at the National Library of Medicine in Bethesda, MD. It consists of over 1,500,000 volumes. Under the Medical Library Assistance Act of 1965 (79 Stat. 1059, 42 U.S.C. 28b) the Congress attempted to make this knowledge available to all "qualified persons" regardless of their geographical location. The following regional medical libraries have been funded to further this purpose. They are 1) The New England

Regional Medical Library Service at the Countway Library of Medicine at Harvard Medical School, 10 Shattuck St., Boston; 2) the New York and New Jersey Regional Medical Library Service at the New York Academy of Medicine, East 103rd St., New York City; 3) the Mid-Eastern Regional Medical Library at the Library of the college of Physicians, 19 South 22 St., Philadelphia; 4) the Mid-Atlantic Regional Medical Library Service of the National Library of Medicine, PO Box 30260, Bethesda, MD; 5) the Kentucky, Ohio, Michigan Regional Library at the Health Sciences Library at Wayne State University, 645 Mullett St., Detroit; 6) the Southwestern Regional Medical Library, A.M. Calhoun Medical Library, Woodrust Research Building, Emory University, Atlanta; 7) the Midwest Regional Medical Library at the John Crerar Library, Chicago; 8) the Pacific Northwest Health Sciences Library at the University of Washington, Seattle; and 9) the Pacific Southwest Regional Medical Library Service at the Biomedical Library of the University of California, Los Angeles.

The statute provides that "qualified persons and organizations shall be entitled to free loan service." Although some of these regional libraries have attempted to restrict the meaning of qualified to other medical libraries, this definition is not consistent with the wording of the statute or its purpose. An excellent argument can be made for the proposition that any individual who can read and who has a desire to research a particular medical question is a "qualified" user under the Act and should be given not only access to the library but free loan privileges as well. This proposition has not yet been argued in court (*see* 42 C.F.R. 59a.31 *et seq.*).

Once you obtain access to the medical library, your main source of information will be the card catalog for books and medical indexes for periodicals. The major index is *Index Medicus,* which is published by the National Institutes of Health (NIH) and indexes about 3000 worldwide medical periodicals under approximately 8000 subject headings. These subject headings are constantly being updated and appear each year as part II of the January issue designated "Medical Subject Headings." Listings are also grouped by the author's last name. In addition to the yearly indexes, there are also a *cumu-*

lative edition and an *abridged* edition of *Index Medicus*. There are also approximately fifteen other more specialized indexes to the medical and biological literature.

Finally, computer searches can locate a complete listing of all articles published over a given period of time on a particular subject matter. Further information on ordering these searches and their cost can be obtained from the reference librarian of your local medical library.

For more specific information on searching the medical literature, *see* Beatty, *Searching the Literature Comes Before Writing the Literature*, 79 Ann. Internal Med. 917–24 (1973).

Index

Abandonment (*see also* Treatment), 50, 57, 77, 78, 81

Abortion (*see also* Consent, informed; Pregnancy and birth), 112, 120–122, 134, 278; pill, 139

Acquired immune deficiency syndrome (AIDS) (*see also* Human immunodeficiency virus), xv, 1, 29, 53, 61, 155, 159, 192, 196, 220, 261, 280

Admission (*see also* Discharge; Transfer), 67–79

AIDS. *See* Acquired immune deficiency syndrome

Alabama, 180

Alcohol (*see also* Drugs), 50, 201, 253

American Association of Homes for the Aging, 35

American Bar Association, 227, 247, 248

American Civil Liberties Union, ix, x, 254, 276

American College of Emergency Physicians, 51

American College of Obstetricians and Gynecologists (ACOG), 125, 127, 135, 139

American College of Physicians, 35

American College of Surgeons, 35, 56, 107

American Hospital Association (AHA), 34–36, 40, 42–44; Special Committee on Biomedical Ethics, 42

American Medical Association (AMA), 24, 36, 76, 177, 222, 223, 227, 240, 247, 248

American Medical Dictionary, 24

American Nurses Association (ANA), 27, 177

American Nursing Home Association, 35

American Psychiatric Association (APA), 186

American Society of Internal Medicine, 72

American Society of Law and Medicine, 276

American Specialty Boards (*see also* Specialists), 23

Amniocentesis. *See* Tests and procedures

Anencephalic newborns, 228

Angela C., case of, 128–130

Arizona, 44, 45, 58, 59, 200

Artificial implants. *See also* Organ transplantation, 147, 221

Assault (*see also* Battery), 85, 124, 190

Atomic Energy Commission. *See* Government agencies

Autopsy (*see also* Death; Organ donation) 233–236

Baby Doe, 212, 214; regulations, 212, 213

Baby Fae, 146, 221

Battery (*see also* Assault), 84, 85, 100, 109, 110, 190

Best Interests Doctrine, 153, 207, 212, 214

Board certification. *See* Specialists

Brain death (*see also* Death), 227, 228, 231, 232

Breach. *See* Malpractice

Breast cancer. *See* Cancer

Burial, 230, 232

Bylaws (*see also* Hospitals), 27

About the Author

George J. Annas, J. D., M. P. H., is the Edward R. Utley Professor of Health Law at Boston University School of Medicine and chief of the health law section at Boston University School of Public Health. An internationally recognized expert on health law and medical ethics, he has taught health law courses in schools of medicine, law, and public health for the past fifteen years. He also writes a regular column for the *Hastings Center Report* and the *American Journal of Public Health* and is editor-in-chief emeritus of *Law, Medicine & Health Care*. He has held a variety of public posts including vice-chairman of the Massachusetts Board of Registration in Medicine, and chairman of the Massachusetts Organ Transplant Task Force. His other books include *The Rights of Doctors, Nurses and Allied Health Professionals* (1981), *Reproductive Genetics and the Law* (1987), and *Judging Medicine* (Humana Press, 1988).

Lightning Source UK Ltd.
Milton Keynes UK
UKOW06f1252161017
311069UK00004B/605/P

9 781461 267430